Principles of Exercise Biochemistry

Medicine and Sport Science

Founder and Editor from 1969 to 1984
E. Jokl, Lexington, Ky.

Vol. 27

Series Editors
M. Hebbelinck, Brussels
R.J. Shephard, Toronto, Ont.

Basel · München · Paris · London · New York · New Delhi · Singapore · Tokyo · Sydney

Principles of
Exercise Biochemistry

Volume Editor
J.R. Poortmans, Brussels

44 figures and 32 tables, 1988

Basel · München · Paris · London · New York · New Delhi · Singapore · Tokyo · Sydney

Medicine and Sport Science

Published on behalf of the Research Committee of the International Council of Sport Sciences and Physical Education

Library of Congress Cataloging-in-Publication Data
Principles of exercise biochemistry.
(Medicine and sport science; vol. 27)
Includes index.
1. Exercise-Physiological aspects. 2. Muscles-Metabolism. 3. Biochemistry.
I. Poortmans, J.R. II. Series.
QP301.P75 1988 612′.04 88-21603
ISBN 3-8055-4790-0

Drug Dosage
The authors and the publisher have exerted every effort to ensure that drug selection and dosage set forth in this text are in accord with current recommendations and practice at the time of publication. However, in view of ongoing research, changes in government regulations, and the constant flow of information relating to drug therapy and drug reactions, the reader is urged to check the package insert for each drug for any change in indications and dosage and for added warnings and precautions. This is particularly important when the recommended agent is a new and/or infrequently employed drug.

Contents

Preface

For millions of years movement has been the only precise way for living creatures to contact one another and share information which has steadily developed former primitive life into social communities. Amongst the fundamental functions involved in these tasks, specialization has been partially oriented to muscle tissue. The latter is the only one capable of increasing its metabolism several hundred times within a matter of seconds. This specific adaptation is found from the invertebrate to the human who has had to use this peculiarity to fight or to escape. For nearly 200 years muscle has been the tool for physiologists and biochemists who have used this tissue to determine the basic mechanisms of membrane properties and energy transduction. For nearly 20 years several textbooks have been devoted to exercise or work physiology in order to communicate the scientific knowledge collected from animal and human physical activities. However, it is rather difficult to find a compilation text on the biochemical effects of exercise since specific subjects are published separately as reviews in journals or yearbooks. Thus, this book is intended for those who wish to have access to applied biochemistry in the field of physical exercise.

The main scope of this book is to give a coordinated coverage of the functional biochemical responses that accompany single and repeated bouts of exercise with major emphasis on humans. Keeping the whole subject within bounds has involved some sacrifices! We have decided to focus the discussion on regulatory mechanisms and new concepts engaged in energy metabolism rather than on detailed metabolic pathways. Thus, the present contribution is directed to undergraduate and graduate students having a basic knowledge of human biochemistry. Specific strategies and adaptations have been developed and some practical implications are given whenever possible.

This book is divided into three parts. Chapters 1–3 deal with the molecular aspects of muscular contraction, fibre type distribution of skeletal muscle and the basic aspects of metabolic regulation. The various authors have tried to provide the reader with new aspects of recent principles on biochemistry which could be applied to exercise conditions. Chapters 4–7 contain information on the utilization of the different fuels such as carbohydrates, purine nucleotides, lipids and proteins. Lastly chapters 8–10 deal with metabolic integration to different types of events, mechanisms of muscular fatigue, and the impact of exercise on metabolic disorders.

We are indebted to students and postdoctoral fellows who have helped us reformulate essential biochemical elements and integrate them with physiological concepts. We are also grateful to S. Karger Co. who have given us encouragement to pursue this task and have assured a high quality of publication.

Brussels, April 1988   *Jacques R. Poortmans*

Poortmans JR (ed): Principles of Exercise Biochemistry.
Med Sport Sci. Basel, Karger, 1988, vol 27, pp 1–21.

Molecular Aspects of Muscular Contraction

Jack A. Rall

Department of Physiology, Ohio State University, Columbus, Ohio, USA

Introduction

Muscle is a machine that converts free energy of adenosine 5′-triphosphate (ATP) into force and/or work and heat. The main objectives of this chapter are to: (a) briefly describe contemporary views concerning mechanisms of activation, contraction and relaxation of muscle, especially skeletal muscle; (b) discuss mechanisms of ATP utilization by muscle during contraction and relaxation, and (c) delineate pathways for immediate supply of energy to the contractile machinery. An introductory understanding of skeletal muscle structure and function is assumed [60].

Sliding Filaments and Muscular Contraction

Contemporary thought on the mechanism of muscle contraction can be traced to 1954 when Huxley and Niedergerke [32] and Huxley and Hanson [37] independently proposed that contraction occurs via sliding filaments. This proposal was based on the light-microscopic observations that the A band of vertebrate skeletal muscle did not change its length, within experimental error, when living fibers [32] or myofibrils from extracted muscle [37] were stretched or shortened actively or passively. Thus, 'over the usual range of physiological shortening' (rest length to 0.65 rest length) [37], muscular contraction appeared to occur by relative sliding of thick and thin filaments without a change in filament length.

Mechanism of Filament Sliding:
Attached Cross-Bridges as Independent Force Generators

Acceptance that contraction occurs via changes in amount of overlap between two sets of interdigitating filaments does not specify a mechanism of contraction. Within the context of sliding filaments, a variety of theories of contraction have been proposed. The most widely accepted view is that filament sliding is due to cyclic reactions consuming ATP between projections on thick myosin-containing filaments, here called cross-bridges, and active sites on thin actin-containing filaments [31, 34]. Cross-bridges are thought to act as independent force generators. The most compelling evidence in support of this proposal is that isometric force production is directly proportional to myofilament overlap when a muscle is stretched from the plateau of the force-length relationship [23]. There are two assumptions underlying this interpretation. First, activation of contraction is maximum and unaltered by stretch. Second, force generation by cross-bridges is unaltered by changes in lateral spacing of thick and thin filaments which occur when a muscle is stretched. The former point is uncontroversial and the latter point is a subject of current investigation.

Another prediction of this theory is that maximum velocity of shortening (V_{max}) is independent of number of attached cross-bridges and thus independent of filament overlap and degree of activation [31]. This prediction has been verified [13]. Assuming that cross-bridges exhibit a finite rate of attachment, the probability of attachment decreases as velocity of filament sliding increases. Thus, force development (dependent on number of attached cross-bridges) decreases with increased velocity of shortening resulting in a hyperbolic force versus velocity curve. At V_{max}, number of attached cross-bridges is at a minimum but not at zero. Even though some cross-bridges are still attaching and pulling, no external force is generated because these cross-bridges are offset by others that have gone through a structural change and are now resisting shortening before detachment. V_{max} results from a balance of force generating and resisting cross-bridges. Therefore, number of attached cross-bridges as determined by degree of activation or filament overlap does not alter V_{max}. Thus, V_{max} reflects speed at which myosin cross-bridges are able to interact with actin components of thin filaments. Support for this view is provided by the observation that V_{max} in a variety of animal species is roughly proportional to their actin-activated myosin ATPase activities in solution [1]. The myosin molecule contains 6 subunits, 2 of high molecular weight called heavy chains

and 4 of low molecular weight called light chains. Myosin exhibits polymorphism [44] in both heavy and light chain components. Attempts are
being made to elucidate which component(s) of the myosin molecule is
responsible for diversity in V_{max} [52].

Chemical and Physical Nature of Cross-Bridges
      Information is available concerning the chemical and physical nature
of cross-bridges. Each myosin molecule is highly asymmetric with: (a) a
long tail which aggregates with tails from other myosin molecules to form
the core of a thick filament containing about 300 myosin molecules, and
(b) two globular heads which form cross-bridges. Each head binds ATP and
actin. A cross-bridge cycle has been proposed in which the cross-bridge
'swings' out variable distances to attach to actin with a particular orientation and then changes its angle of attachment, during the working stroke,
so that force generation and relative filament sliding of about 10 nm occurs
while ATP is hydrolyzed [35]. This model was based primarily on structural considerations as determined from electron microscopy and X-ray
diffraction. This model of cross-bridge function requires that myosin
exhibit two sites of flexibility, one which allows the cross-bridge to swing
out and the other which allows a change of attachment angle. The isolated
myosin molecule exhibits the requisite flexibility [24]. It was suggested that
the attachment angle of the cross-bridge was 90° and the angle at the end of
the working stroke was 45°. These angles corresponded to measured angles
of cross-bridges in muscle at rest and during rigor where actin and myosin
are irreversibly linked [35].
      Recent structural evidence requires modification of the above model.
It has not been possible to confirm the 90–45° cross-bridge rotation during
the working stroke in contracting muscle. Rather, probes (paramagnetic or
fluorescent) attached to the myosin head do not change orientation during
contraction [7]. Thus, at least a portion of the myosin head does not rotate
during the working stroke. It is still possible that the part of the myosin
head which changes orientation during the working stroke has not been
labelled with probes. Nonetheless, the notion that the whole cross-bridge
changes its angle during contraction is unverified. Furthermore, significance of the double-headed nature of the myosin molecule is a mystery.
      Other structural and mechanical evidence has confirmed and ex-
tended the cross-bridge model of contraction. Recent advances in high
intensity X-ray sources and detectors have allowed structural changes during contraction to be monitored with millisecond resolution [40]. In gen-

eral, structural changes precede force generation and lag force decline during relaxation. These studies are consistent with the necessary condition that a cross-bridge 'swings' out to a thin filament *before* force is developed. These results are in agreement with mechanical experiments where stiffness of a muscle fiber was taken to reflect number of attached cross-bridges. Stiffness precedes force development [5], indicating that there is a delay between cross-bridge attachment and force generation. Also, this evidence suggests that there are multiple attached cross-bridge states. Furthermore, cross-bridges are thought to contain an instantaneous elasticity whose precise location is unknown [33, 57]. The cross-bridge elasticity represents about 50% of the total series elasticity of a muscle with the remainder residing in tendons.

Myosin and Actomyosin ATPase in Solution:
Relation to Cross-Bridge Cycle

Free energy of ATP provides the driving force for cross-bridge cycling. Lymn and Taylor [45] have provided the framework for the contemporary view of kinetic mechanisms of ATP hydrolysis by myosin and by actin + myosin (actomyosin) in solution. Consider splitting of ATP by myosin. Myosin is usually studied in solution as soluble fragments, heavy meromyosin or subfragment-1, which contain the ATP hydrolysis and actin-binding properties of the whole molecule. Myosin (M) is an ATPase which hydrolyzes Mg.ATP at a slow rate. Binding of ATP to myosin (M.ATP) and its hydrolysis to adenosine 5′-diphosphate (ADP) and inorganic phosphate (Pi) on myosin (M.ADP.Pi) are rapid processes but dissociation of products from myosin is slow and thus the rate limiting step in the overall cycle. The predominant form of myosin in solution and of cross-bridges in resting muscle is a myosin-product complex (M.ADP.Pi). Even though ATP has been hydrolyzed, this complex stores free energy of ATP. Actin greatly accelerates myosin ATPase activity by binding to the myosin-product complex (A.M.ADP.Pi) and increasing rate of product dissociation. ATP binds to the actomyosin complex (A.M) causing dissociation of actin and myosin. ATP is subsequently hydrolyzed on free myosin.

An attraction of this proposal is that it fits in a natural way with accepted views of the cross-bridge cycle in muscle. In each cycle in solution, one ATP is split and actin and myosin go through an association and disassociation cycle analogous to the cross-bridge attachment and detachment cycle in muscle (fig. 1). The working stroke of the cross-bridge cycle where free energy of ATP is converted into mechanical energy is tenta-

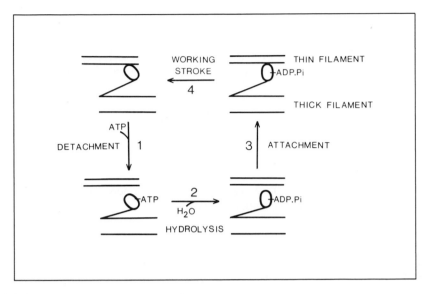

Fig. 1. Cross-bridge cycle and its relation to actomyosin ATPase based on a model proposed by Lymn and Taylor [45]. Changes shown in cross-bridge orientation during the cycle, based on 35, are for purposes of illustration only. Actual physical orientations of the cross-bridge during the cycle are a matter of speculation [7].

tively identified as the release of products (ADP and Pi, step 4) because this step is associated with a large drop in free energy in solution. In this scheme ATP is hydrolyzed when actin and myosin are disassociated and not at the time of external work production.

More recent information has led to expansion of the Lymn-Taylor scheme to include additional states [21]. A major modification is inclusion of an additional actomyosin state (A.M′.ADP) following release of Pi from A.M.ADP.Pi. Also, it is kinetically possible to have a nondissociating pathway for ATP hydrolysis, i.e. it is not obligatory that actin and myosin dissociate and reassociate for each ATP hydrolyzed. This pathway is favored at low ionic strength. Its importance at physiological ionic strength is questionable. These modifications are described in figure 2.

Whereas major advances have occurred in the understanding of the kinetics of actomyosin interaction, it is important to realize that solution studies provide only a limited model for muscle contraction. These studies lack: (a) constraints of the myofilament lattice and concomitant stresses

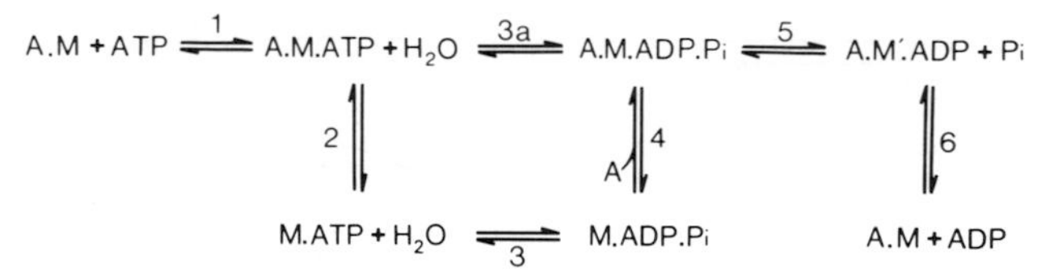

Fig. 2. Actomyosin ATPase mechanism including a nondissociating pathway for ATP hydrolysis (step 3a) and multiple actomyosin states.

and strains thereof, and (b) actual conversion of free energy into force and/or work. Thus, solution kinetic studies are not analogous to an isometric contraction. Nor are solution studies analogous to shortening under zero force, i.e. shortening at V_{max}, since V_{max} is a result of a balance of forces among cross-bridges (see above).

Actomyosin ATPase in Structured Systems

Techniques have been developed so that kinetics of actin and myosin interaction can be studied in relation to force generation and/or shortening [21]. The most useful preparation is the 'skinned' fiber which consists of demembranated muscle fiber(s). The membrane is removed by mechanical [50] or chemical means (e.g. glycerol extraction). Myofilaments are intact and the fiber is capable of generating normal steady state force and shortening velocity in response to constituents in the medium. An important limitation of skinned fiber preparations has been time delays due to diffusion of constituents of the medium into myofilaments. These delays distort and limit temporal resolution of both mechanical and chemical measurements compared to those done on intact fibers and on proteins in solution. A recent advance has been development of a technique to greatly reduce diffusional delays in skinned preparations. This is accomplished by exposing a preparation in rigor (i.e. no ATP, cross-bridges attached) to a chemically inert precursor of ATP, P^3-1-(2-nitro)phenylethyl adenosine 5′-triphosphate or 'caged ATP', which is photolabile [21, 39]. Once caged ATP is distributed throughout the preparation, ATP is released by action of a laser flash on the inert precursor. This rapid release avoids major delays in activating skinned fibers. Furthermore, a controlled and known amount of ATP (or Pi or ADP) can be released and effects on fiber force production and stiffness measured.

Using this approach kinetics of cross-bridge detachment and attachment have been studied [22]. One important conclusion is that the predominant intermediates present during contraction are attached states, since detachment, hydrolysis and reattachment with force generation are all much more rapid than the overall cycling rate. Thus, the rate-limiting step in the cycle occurs during a cross-bridge attachment state. Furthermore, the step of the cycle leading to a force-generating state is thought to be release of Pi (i.e. transition from A.M.ADP.Pi to A.M'.ADP) [26]. This implies that A.M.ADP.Pi is an attached but non-force-generating state. This result is consistent with the observation that increases in muscle stiffness precede force development [5]. Confirmation of the role of Pi comes from experiments where an increase in Pi results in decreased force production and a less than proportional decrease in muscle stiffness. This result suggests that Pi binding to A.M.'ADP forms A.M.ADP.Pi with a concomitant decrease in force but not a proportional decrease in stiffness since this is still an attached state. The effect of Pi might also help explain the decline of force with prolonged contraction.

New approaches to the study of muscle function promise exciting advances. These approaches include: crystallization of actin and myosin (subfragment-1) so that structure of actin and myosin interaction may be deduced at an atomic level, utilization of molecular biology techniques to elucidate function and control of myosin heavy and light chain multi-gene families [2, 38], use of monoclonal antibodies to probe contractile protein function, and techniques to exchange contractile protein subunits in skinned fibers in order to study function of individual subunits [49].

Activation of Muscle Contraction

It is established that calcium is the intracellular activator of contraction in all types of muscle. Heilbrunn was a pioneer in supporting the belief that calcium mediated many intracellular processes [4]. The contemporary view of the role of calcium in muscular activation and relaxation can be traced to biochemical experiments of Ebashi and also Weber in the late 1950s [see 12 for historical perspective]. These investigators realized the importance of micromolar calcium concentrations in activation of acto-myosin ATPase under physiological conditions. Ebashi advanced the understanding of the mechanism of activation when he discovered and elucidated the role of the first known calcium-binding protein, troponin.

In skeletal muscle the action potential propagates through invaginations of the surface membrane called transverse tubules (T-tubules) and the accompanying depolarization causes release of intracellular calcium stored in terminal cisternae of the sarcoplasmic reticulum (SR). SR and T-tubules are separated at closest approximation by a gap of 15 nm. This gap is bridged by connecting structures. The molecular mechanism of calcium release is not known with certainty though intramembrane charge movement in T-tubules appears to be coupled to calcium release by SR [54]. Change movement may trigger calcium release: (a) directly; (b) indirectly through a chemical messenger like inositol trisphosphate, or (c) indirectly by gating a calcium channel in the T-tubules which induces calcium release. This last possibility, calcium-induced calcium release, may be operative in heart [17]. For a recent review of this active research area, see Endo [16].

An important advance in studying intracellular calcium regulation has been development of probes that signal free calcium concentration in living cells in the submicromolar range and that follow changes in calcium concentration on a millisecond time scale [for an evaluation of techniques, see 3]. In skeletal muscle with stimulation calcium diffuses into the sarcoplasm from SR and raises free calcium concentration from approximately 100 nM to 10 μM during a tetanus.

During activation calcium binds to thin filaments, specifically to calcium-specific sites on the troponin C subunit of the troponin molecule and this leads to activation of cross-bridge cycling and contraction. Another component of the thin filament is tropomyosin, long helically shaped molecules that attach end-to-end and lay in the two grooves of the actin double helix. Stoichiometry is such that 1 troponin is associated with 1 tropomyosin and 7 actins. It has been proposed that at rest tropomyosin physically blocks myosin from reaching the actin-binding site necessary for contraction and that during contraction tropomyosin 'slides' toward the center of the groove by about 2 nm allowing actin and myosin interaction [25, 36]. This 'steric blocking' model of muscular activation predicts that, at rest, myosin would be unable to bind to actin. On the contrary, recent biochemical evidence suggests that myosin binding to the actin + tropomyosin + troponin complex is independent of calcium ion concentration but nevertheless actomyosin ATPase activity is controlled by calcium [6]. These observations are inconsistent with the simple steric blocking model and imply that tropomyosin-troponin inhibits a kinetic step in ATP hydrolysis after attachment, possibly transition from A.M.ADP.Pi to A.M'.ADP (see above). The observed temporal sequence of structural changes during muscular activation is consistent with an inhibitory role of tropomyosin since tropomyosin moves before cross-bridges reach thin filaments and both events occur before force development [40]. Other recent experiments suggest that there may be long range molecular cooperativity within the thin filament during activation. This cooperativity implies that calcium binding to one troponin C turns on more than 7 actins [48].

Only thin-filament activation has been considered but other modes of activation including thick-filament activation exist in vertebrate smooth muscle and in some invertebrate muscles. These and other pertinent topics are discussed in a recent monograph which emphasizes comparative approaches to study of muscle activation [53].

Relaxation from Muscular Contraction

SR stores calcium at rest, releases it during activity and sequesters calcium by an ATP-dependent mechanism to cause muscular relaxation [19]. SR is an enclosed intracellular reticular structure that forms a network surrounding each myofibril in skeletal muscle. The SR membrane contains a calcium pump protein that is uniformly distributed along longitudinal tubules and terminal cisternae. SR contains the calcium-binding protein calsequestrin, located within the lumen of the terminal cisternae, which stores calcium, thus reducing the gradient against which the pump must work.

Investigation of the molecular mechanism of calcium accumulation by SR has been greatly facilitated by development of the isolated SR vesicle preparation [47]. These fragmented SR vesicles have a diameter of about 100 nm with outer surface of vesicles corresponding to surface of SR that faces the sarcoplasm in vivo. Up to 90% of the membrane protein of these vesicles is attributed to the calcium pump. This pump is also a Ca^{2+}-transport ATPase which is activated by submicromolar concentrations of calcium. Steps involved in accumulation of calcium by SR include: (a) binding of 1 mol of Mg.ATP and 2 mol of calcium per mol of transport protein (E); (b) rapid generation of a high-energy phosphorylated interme-

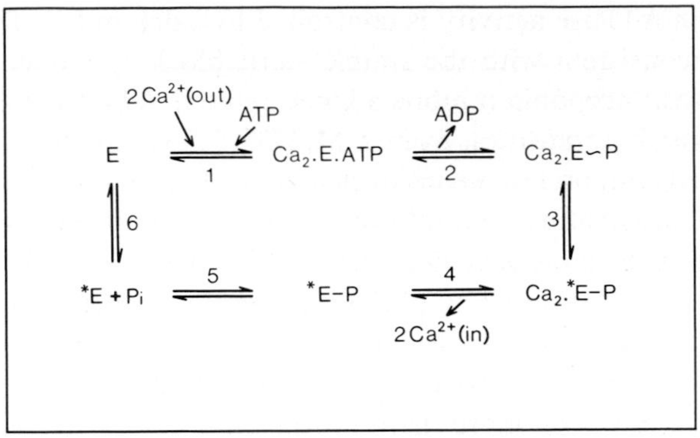

2 Ca^{2+}(out)

E $\rightleftharpoons$ Ca$_2$.E.ATP $\rightleftharpoons$ Ca$_2$.E~P

ATP   1   2   ADP

6   3

*E + P$_i$ $\rightleftharpoons$ *E-P $\rightleftharpoons$ Ca$_2$.*E-P

5   4

2 Ca^{2+}(in)

Fig. 3. Reaction scheme for ATP hydrolysis coupled to calcium transport by calcium pump of sarcoplasmic reticulum. E~P has a high affinity for calcium and *E-P has a low affinity for calcium.

diate (Ca$_2$·E~P); (c) calcium translocation through membrane; (d) discharge of calcium into SR lumen after a dramatic loss of affinity during translocation, and (e) decomposition of *E-P and recycling of enzyme. One cycle of the ATPase couples hydrolysis of 1 ATP to uptake of 2 calcium. The free energy of ATP hydrolysis is used to reduce affinity and allow dissociation of the calcium-enzyme complex inside SR vesicles where calcium is higher than outside. These steps are summarized in figure 3.

Polymorphism of the Ca^{2+}-transport ATPase has been demonstrated with mammalian fast-twitch muscle displaying faster elementary reaction steps than slow-twitch muscle [10]. Also, the concentration of calcium-transport sites appears to be higher in SR of fast-twitch than slow-twitch muscle. These factors along with the higher volume of SR in fast-twitch muscle [14] are probably responsible for the faster rate of mechanical relaxation in fast-twitch than in slow-twitch muscle (though see below).

There has been doubt expressed as to whether the rate of calcium uptake by isolated SR vesicles is rapid enough to explain the rate of muscular relaxation. This uncertainty may be due to one or more of the following: (a) SR transport properties may be modified by isolation; (b) other calcium-binding proteins may supplement SR function in vivo, and/or (c) relaxation may be influenced by mechanical conditions along the fiber, such as asynchrony of calcium uptake, causing nonhomogeneity of force production during relaxation [19]. Experiments employing electron probe

X-ray microanalysis to measure total calcium concentration in individual SR terminal cisterna at rest, during prolonged contraction, and just after relaxation have determined that SR takes up, during relaxation, an amount of calcium equivalent to that bound to troponin [56]. Thus, under conditions of prolonged contraction, the SR calcium pump is sufficiently rapid to cause relaxation.

Nonetheless, muscles that contract for brief periods of time relax up to 3 times faster than for prolonged contractions. It has been suggested that this increment in relaxation rate may be due to a soluble calcium-binding protein called parvalbumin [20]. Parvalbumin is found in high concentration in fish, sometimes 10 times higher than troponin, and in amphibian skeletal muscle. It is also in higher concentration in fast-twitch than in slow-twitch mammalian muscle. It has been suggested that parvalbumin facilitates relaxation during brief contractions but during prolonged contractions it becomes saturated with calcium and then relaxation is attributable to SR alone. Calcium bound to parvalbumin during contraction is thought to slowly return to SR after relaxation. This has been verified [56]. Thus, at the end of relaxation not all of the calcium released during contraction has returned to SR. The notion of parvalbumin as a 'soluble relaxing' factor is still a matter of speculation.

Energetics of Muscular Contraction

Energetics of muscular contraction can be considered from two points of view. The first deals with utilization of chemical energy by contractile machinery during contraction and relaxation. The second considers aerobic and anaerobic metabolic processes replenishing utilized chemical energy. In the steady state, rate of chemical energy utilization by contractile machinery equals rate of chemical energy replenishment by metabolic machinery. Only chemical energy utilization by contractile machinery in isolated muscles and fibers will be considered in detail. Metabolic processes replenishing utilized chemical energy are considered in depth in other chapters.

During contraction and relaxation of intact muscle, ATP is utilized by cross-bridges (actomyosin ATPase) and the calcium pump (Ca^{2+}-transport ATPase). In the absence of metabolic inhibitors, ATP hydrolysis is not observable. What is observed is splitting of phosphocreatine (PCr) to creatine (Cr) and inorganic phosphate (Pi). This is true because the enzyme

ATP:creatine phosphokinase rapidly catalyzes a reversible reaction that leads to rephosphorylation of ADP that is formed. These reactions are summarized as:

$$ATP + H_2O \rightarrow ADP + Pi, \tag{1}$$
$$\underline{ADP + PCr \rightleftharpoons Cr \quad + ATP,} \tag{2}$$
$$PCr \rightarrow Cr \quad + Pi. \tag{3}$$

ATP levels are well buffered during contraction. For example, when 90% of the total PCr is split in human muscle during contraction, there may be only a 10% decrease in ATP levels [30].

Energetics of Isometric Contractions

Much of the current understanding of skeletal muscle energetics is based on studies by Hill [27] over a period of more than 50 years. Best understood is frog skeletal muscle studied at 0 °C. Under these conditions there is a complete separation of chemical energy utilization during contraction due to initial processes from chemical energy replenishment after contraction due to recovery processes. Rate of ATP hydrolysis during an isometric tetanus of isolated frog skeletal muscle is initially high during force development and then decreases by 30% to a steady rate during force maintenance [28]. This steady rate of ATP hydrolysis during contraction is about 500 times greater than the resting rate of ATP utilization [51]. Since hydrolysis of ATP (and net PC splitting) is an exothermic or heat-producing reaction, steady rate of ATP hydrolysis is reflected in a steady rate of energy (heat + work) liberation during an isometric tetanus [28]. Interestingly, more energy is liberated during the initial part of an isometric tetanus than can be attributed to ATP hydrolysis alone. Other heat-producing reactions must occur during this time [61].

How are steady ATP hydrolysis and energy liberation rates during contraction in intact muscle partitioned among major ATP-utilizing processes? These processes include actomyosin ATPase, Ca^{2+}-transport ATPase and Na^+,K^+-ATPase. The Na^+,K^+-ATPase does not contribute significantly to ATP utilization during contraction because muscles soaked in strongly hypertonic solutions still generate and propagate action potentials but do not produce force or liberate energy [55]. How then is ATP hydrolysis during contraction partitioned between cross-bridge cycling and calcium pumping? An approach to this question is derived from consideration of the sliding filament-attached cross-bridge model of contraction and the length-tension relationship [23]. Maximum isometric force occurs at sarcomere lengths of 2.0–2.25 μm where overlap of myofilaments is optimum. If a resting muscle is stretched to sarcomere lengths greater than 3.65 μm and then stimulated, force production is greatly depressed since myofilament overlap is near zero. Thus, actomyosin ATPase is inhibited by preventing actin and myosin interaction. Nonetheless, about 30% of energy liberation and ATP hydrolysis observed at maximum myofilament overlap still is observed at no overlap [29, 55]. This fraction is essentially the same for tetanic contractions of varying durations. Similar results have been observed in mammalian fast- and slow-twitch skeletal muscles [8]. Since calcium is still released at zero myofilament overlap, these results have been interpreted to mean that about 30% of ATP utilized during contraction is utilized by the calcium pump. Therefore, 70% is utilized by cross-bridges in maintenance of force irrespective of contraction duration or muscular fiber type. This interpretation suggests two conclusions: (a) the calcium pump utilizes ATP continually during a tetanus and thus calcium cycles continuously, and (b) rate of ATP hydrolysis by cross-bridges is 'matched' to that of the calcium pump such that the ratio of the two remains constant in different muscular types. This matching is reasonable since it would be pointless to have a fast contracting muscle relax slowly because it would not be able to contract rapidly a second time until it slowly relaxed from the first contraction.

Economy and Power Output during Contraction

Relationship of isometric force production to metabolic cost of contraction is important in characterizing energetic properties of muscles. The ratio of isometric force to metabolic cost is called economy of contraction [51]. Since maximum force production per cross-sectional area varies little in different muscles, the major determinant of economy is metabolic cost reflected in steady rate of ATP hydrolysis or energy liberation. Thus, economy of contraction is inversely proportional to steady rate of ATP hydrolysis or energy liberation. Economy of contraction varies remarkably across the animal kingdom [51]. Examples are shown in table I where economy is expressed relative to that observed for frog sartorius muscle at 20 °C. The range is enormous. For example, to maintain comparable force, frog sartorius muscle would utilize as much ATP in 1 s as the anterior byssus retractor muscle of bivalve molluscs (ABRM) would utilize in 46 min in the tonic state. Mammalian skeletal muscle has not been extensively studied in this regard and variation in economy listed in table I certainly does not represent the whole range. Molecular sources of this

Table I. Variation in economy of force maintenance during isometric contraction

	Economy relative to frog sartorius[1]
Frog sartorius	1
Tortoise skeletal muscle	40
Mammalian skeletal muscle (mouse)	
Extensor digitorum longus	1
Soleus	3.4
Mammalian smooth muscle (rabbit)	100
Invertebrate smooth muscle (ABRM, mollusc)	
Phasic	250
Tonic	2,750

[1] Data taken from studies sited in Rall [51].

variation are due to differences in rate of ATP utilization by cross-bridges and the calcium pump. Thus, economy is inversely proportional to actomyosin ATPase and Ca^{2+}-transport ATPase.

Economy of contraction is related to maximum power output of a muscle. Power is the rate of doing work and is equal to force times velocity of shortening. Empirically, maximum power output of a muscle is roughly equal to $0.1 \times V_{max} \times F_{max}$, where F_{max} is maximum isometric force production per cross-sectional area. Since F_{max} varies little in different muscles, maximum power output is determined primarily by V_{max}. Since V_{max} is roughly proportional to rate of actomyosin ATPase [1], maximum power output is proportional to actomyosin ATPase. Thus, there is a reciprocal relationship between economy and maximum power output of a muscle. These relationships are summarized in table II.

The major conclusion is that there is a compromise in muscle design such that a muscle cannot be highly economical in force maintenance and also contract with a high power output.

Flexibility is incorporated into a muscle in vivo by existence of fibers with varying economies of contraction (and thus power outputs) and by capability of differential fiber activation by nerves. Fibers can be recruited in a way to vary economy and power output of a muscle to appropriately match demand. In mammalian skeletal muscle the concept of fiber types is

Table II. Compromises in muscular design

Economy	$\propto$	1/ A.M ATPase
V_{max}	$\propto$	A.M. ATPase
Maximum power	=	$0.1(F_{max})(V_{max})$
	$\propto$	A.M ATPase
$\therefore$ economy	$\propto$	1/ maximum power

well established and is the subject of a later chapter. The iliofibularis muscle of the South African clawed toad is a striking example of intramuscular diversity. In this skeletal muscle at least 5 more or less distinct fiber types have been identified [15]. When individual fibers are studied, economy of contraction (and maximum power output) varies by 17-fold among fiber types [15]. This provides a large range of flexibility to deal with different motor tasks. Another observation is that even within a defined fiber type there may be a 2-fold variation in economy. This result suggests that fibers within a fiber type are not homogeneous and raises the possibility that there may be a continuum of fibers rather than discrete fiber types.

Energetics of Muscles that Shorten or Are Stretched during Contraction

When a muscle shortens and does work, there is a large increase in rate of ATP hydrolysis and energy liberation [18, 42]. This phenomenon was first observed by Fenn [18] who measured the rate of energy liberation in isometric and isotonic contractions. The increase in ATP hydrolysis and energy liberation during working contractions is thus known as the Fenn effect. Since the rate of ATP hydrolysis in a maximally working contraction can be 4 times greater than steady rate of ATP hydrolysis in an isometric tetanus [42], the increase in rate of ATP hydrolysis above rest is about 2,000-fold in the working contraction. In terms of the sliding filament-attached cross-bridge theory, the increase in ATP hydrolysis rate during maximum work production is due to more rapid cycling of cross-bridges [31, 33]. Thus, a step in the cross-bridge cycle, possibly related to product release from actomyosin, is sensitive to force production and shortening.

During many movements in vivo, muscles are forcibly lengthened (stretched) during contraction. If an isolated skeletal muscle is stretched during contraction, it responds by resisting stretch with a force greater than the isometric value. At the same time, rate of ATP hydrolysis is greatly decreased compared to that in an isometric contraction [9]. Stretch of a muscle during contraction is resisted by attached cross-bridges and the extra force suppresses the normal cross-bridge cycling rate leading to diminished ATP hydrolysis. Thus, this mode of contraction leads to large force development and requires considerably less ATP than isometric or isotonic contractions. An important mechanism in minimizing energetic cost of locomotion is transient storage of elastic energy in and subsequent recovery of elastic energy from stretched muscles and tendons [11, 46]. About half of the positive work expended during locomotion in humans can be attributed to recovered elasticity of muscles and tendons [46]. Elastic energy of stretch is stored not only in tendons but also in cross-bridges of muscle since cross-bridges are known to possess elastic properties [33].

Efficiency of muscular contraction provides a measure of energetic cost of work production comparable to economy of contraction which provides a measure of energetic cost of force maintenance [51, 59]. Efficiency can be thought of as work (W) actually obtained from a process divided by maximum work (W_{max}) obtainable. The W_{max} obtainable from a reaction is equal to the free energy change (ΔG) for that reaction. Experimentally, W_{max} has been estimated in different ways [51]. Maximum efficiency observed in isolated frog skeletal muscle for a complete contraction-relaxation-recovery cycle (including initial and recovery processes) is about 0.2. Maximum efficiency for a contraction-relaxation cycle is about 0.45. Higher efficiencies can be obtained when muscles are stretched during contraction and then released to do work. Under these conditions elastic energy stored in the muscle and tendons is converted to work during shortening. Efficiency of recovery processes is approximately 0.5, i.e. 50% of free energy of carbohydrate is converted into ATP during recovery and 50% is lost to the environment as heat. Maximum efficiency of chemomechanical energy conversion by cross-bridges must be greater than 0.45 since about 30% of ATP utilized during contraction is utilized by the calcium pump and thus not directly converted to external work [29, 55]. Efficiency of contraction is a function of force (load) against which the muscle shortens. Maximum efficiency of work production occurs when force is about 50% of maximum isometric force. Efficiency of contraction varies little across the animal kingdom when compared to economy of contraction [51]. In general, efficiency bears no relationship to V_{max} or maximum power output contrary to the situation for economy though this point needs further investigation [51].

Reactions Occurring during Intense or Prolonged Muscle Activity

During muscular contraction ATP utilization is replenished by splitting of PCr. During brief contractions, the net reaction is splitting of PCr to Cr and Pi. To clarify what is meant by brief contractions, consider an example. Isometric contractions of human quadriceps femoris muscles (containing approximately 50% slow-twitch fibers) were elicited by electrical stimulation so that initial force production was 50–75% of maximum voluntary force production. Under these conditions there is only enough ATP in the resting muscles to support about 5 s of stimulation and PCr to support about 14 further seconds of stimulation [30]. In this example glycolysis was activated within 5 s of initiation of contraction. Thus, for intense or prolonged activity other reactions must be activated [41]. The first of these reactions is catalyzed by ATP:AMP phosphoryltransferase (also known as adenylate kinase and myokinase):

$$2\,ADP \rightleftharpoons ATP + AMP. \tag{4}$$

Rate of ATP formation by this reaction can be about one-third the maximum rate of the creatine phosphokinase reaction (reaction 2 above) under optimal conditions. This reaction is driven to a greater extent under conditions which lead to removal of AMP by deamination. The irreversible deamination of AMP is catalyzed by adenosine 5′-monophosphate deaminase (AMP deaminase):

$$AMP + H_2O + H^+ \rightarrow IMP + NH_4^+, \tag{5}$$

where IMP is inosine 5′-monophosphate and NH_4^+ is ammonium ion. This reaction proceeds to a significant extent only during intense activity and is probably activated to an important extent by intracellular acidosis. A key factor in activation of the AMP deaminase reaction appears to be the relationship of metabolic stress (or severity of contraction) to aerobic capability of the muscle [58]. For example, during intense contractile activity AMP deaminase reaction is activated to the greatest extent in fast-twitch muscle with a low oxidative capacity and not activated to any appreciable extent in slow-twitch muscle with a high oxidative capacity. During intense contractile activity in fast-twitch skeletal muscle, ATP content can decrease by 50%. Removal of AMP by deamination facilitates the adenylate kinase reaction which in turn limits the decrease in the ratio of ATP to ADP during intense activity. This ratio is important in determining free energy available from ATP hydrolysis. The AMP deaminase reaction may also be important in control of glycolysis since ammonium ions activate phosphofructokinase and IMP may be involved in regulation of phosphorylase activity. Thus, during intense contractile activity in fast-twitch skeletal muscle, the total adenine nucleotide pool (ATP + ADP + AMP) is decreased and IMP and ammonium ions are produced.

Although the AMP deaminase reaction is irreversible, IMP can be reaminated by a reaction sequence involving guanosine tri- and diphosphate (GTP and GDP). This sequence is catalyzed by the enzymes adenylosuccinate synthetase and adenylosuccinate lyase:

$$IMP + aspartate + GTP \rightarrow adenylosuccinate + GDP + Pi, \tag{6}$$

$$adenylosuccinate \rightarrow AMP + fumarate. \tag{7}$$

These reactions plus the AMP deaminase reaction constitute the purine nucleotide cycle [43]. One complete revolution of the purine nucleotide cycle results in net formation of fumarate, GDP, Pi and ammonium ion. IMP reamination is a slow process and does not occur to a significant extent during the contraction period but rather is a part of recovery. The fumarate produced may enter the mitochondria, thereby expanding the pool of citric acid cycle intermediates and promoting mitochondrial respiration. Other recovery processes, including glycolysis and oxidative metabolism, are considered in other chapters.

References

1   Bárány, M.: ATPase activity of myosin correlated with speed of muscle shortening. J. gen. Physiol. 50: 197–216 (1967).
2   Barton, P.J.; Buckingham, M.E.: The myosin alkali light chain proteins and their genes. Biochem. J. 231: 249–261 (1985).
3   Blinks, J.R.; Wier, W.G.; Hess, P.; Prendergast, F.G.: Measurement of Ca2+ in living cells. Prog. Biophys. mol. Biol. 40: 1–114 (1982).
4   Campbell, A.K.: Lewis Victor Heilbrunn, pioneer of calcium as an intracellular regulator. Cell Calcium 7: 287–296 (1986).
5   Cecchi, G.; Griffiths, P.J.; Taylor, S.: Muscular contraction: kinetics of crossbridge attachment studied by high-frequency stiffness measurements. Science 217: 70–73 (1982).
6   Chalovich, J.M.; Eisenberg, E.: Inhibition of actomyosin ATPase activity by troponin-tropomyosin without blocking the binding of myosin to actin. J. biol. Chem. 257: 2432–2437 (1982).
7   Cooke, R.: The mechanism of muscle contraction. CRC crit. Rev. Biochem. 21: 53–118 (1986).
8   Crow, M.T.; Kushmerick, M.J.: Correlated reduction of velocity of shortening and

the rate of energy utilization in mouse fast-twitch muscle during a continuous tetanus. J. gen. Physiol. *82:* 703–720 (1983).

9   Curtin, N.A.; Davies, R.E.: Very high tension with very little ATP breakdown by active skeletal muscle. J. Mechanochem. Cell Motil. *3:* 147–154 (1975).

10   Damiani, E.; Betto, R.; Salvatori, S.; Volpe, P.; Salviati, G.; Margreth, A.: Polymorphism of sarcoplasmic-reticulum adenosine triphosphatase of rabbit skeletal muscle. Biochem. J. *197:* 245–248 (1981).

11   Dawson, T.J.; Taylor, C.R.: Energetic cost of locomotion in kangaroos. Nature, Lond. *246:* 313–314 (1973).

12   Ebashi, S.: Regulation of muscle contraction. Proc. R. Soc. Lond. B *207:* 259–286 (1980).

13   Edman, K.A.P.: The velocity of unloaded shortening and its relation to sarcomere length and isometric force in vertebrate muscle fibers. J. Physiol. *291:* 143–159 (1979).

14   Eisenberg, B.R.: Quantitative ultrastructure of mammalian skeletal muscle; in Peachey, Adrian, Geiger, Handbook of physiology, sect. 10: skeletal muscle, pp. 73–112 (American Physiological Society, Bethesda 1983).

15   Elzinga, G.; Lannergren, J.; Stienen, G.J.M.: Stable maintenance heat rate related to contractile properties of different single muscle fibres from *Xenopus* at 20 °C. J. Physiol. *393:* 399–412 (1987).

16   Endo, M.: Calcium release from sarcoplasmic reticulum; in Shamoo, Current topics in membranes and transport, vol. 25, pp. 181–230 (Academic Press, New York 1985).

17   Fabiato, A.; Fabiato, F.: Excitation-contraction coupling of isolated cardiac fibers with disrupted or closed sarcolemmas. Calcium-dependent cyclic and tonic contractions. Circulation Res. *31:* 293–307 (1972).

18   Fenn, W.O.: A quantitative comparison between the energy liberated and the work performed by the isolated sartorius muscle of the frog. J. Physiol. *58:* 175–203 (1923).

19   Gillis, J.M.: Relaxation of vertebrate skeletal muscle. A synthesis of the biochemical and physiological approaches. Biochim. biophys. Acta *811:* 97–145 (1985).

20   Gillis, J.M.; Thomason, D.; Lefevre, J.; Kretsinger, R.H.: Parvalbumins and muscle relaxation. A computer simulation study. J. Muscle Res. Cell Motil. *3:* 377–398 (1982).

21   Goldman, Y.E.: Kinetics of the actomyosin ATPase in muscle fibers. A. Rev. Physiol. *49:* 637–654 (1987).

22   Goldman, Y.E.; Hibberd, M.G.; McCray, J.A.; Trentham, D.R.: Relaxation of muscle fibres by photolysis of caged ATP. Nature, Lond. *300:* 701–705 (1982).

23   Gordon, A.M.; Huxley, A.F.; Julian, F.J.: The variation in isometric tension with sarcomere length in vertebrate muscle fibers. J. Physiol. *184:* 170–192 (1966).

24   Harvey, S.C.; Cheung, H.C.: Myosin flexibility; in Dowben, Shay, Cell and muscle motility, vol. 2, pp. 279–302 (Plenum Press, New York 1982).

25   Haselgrove, J.C.: X-ray evidence for a conformational change in the actin containing filaments of vertebrate striated muscle. Cold Spring Harb. Symp. quant. Biol. *37:* 341–352 (1973).

26   Hibberd, M.G.; Dantzig, J.A.; Trentham, D.R.; Goldman, Y.E.: Phosphate release and force generation in skeletal muscle fibers. Science *228:* 1317–1319 (1985).

27   Hill, A.V.: Trails and trials, in physiology, pp. 1–374 (Williams & Wilkins, Baltimore 1965).

28   Homsher, E.; Kean, C.J.; Wallner A.; Garibian-Sarian, V.: The time-course of energy balance in an isometric tetanus. J. gen. Physiol. *73:* 553–567 (1979).

29   Homsher, E.; Mommaerts, W.F.H.M.; Ricchiuti, N.V.; Wallner, A.: Activation heat, activation metabolism and tension-related heat in frog semitendinosus muscles. J. Physiol. *220:* 601–625 (1972).

30   Hultman, E.; Sjoholm, H.: Energy metabolism and contraction force of human skeletal muscle in situ during electrical stimulation. J. Physiol. *345:* 525–532 (1983).

31   Huxley, A.F.: Muscle structure and theories of contraction. Prog. Biophys. biophys. Chem. *7:* 255–318 (1957).

32   Huxley, A.F.; Niedergerke, R.: Structural changes in muscle during contraction. Nature, Lond. *173:* 971–973 (1954).

33   Huxley, A.F.; Simmons, R.M.: Proposed mechanism of force generation in striated muscle. Nature, Lond. *233:* 533–538 (1971).

34   Huxley, H.E.: The double array of filaments in cross-striated muscle. J. biophys. biochem. Cytol. *3:* 631–648 (1957).

35   Huxley, H.E.: The mechanism of muscular contraction. Science *164:* 1356–1366 (1969).

36   Huxley, H.E.: Structural changes in the actin and myosin containing filaments during contraction. Cold Spring Harb. Symp. quant. Biol. *37:* 361–376 (1973).

37   Huxley, H.E.; Hanson, J.: Changes in the cross-striations of muscle during contraction and stretch and their structural interpretation. Nature, Lond. *173:* 973–976 (1954).

38   Izumo, S.; Nadal-Ginard, B.; Mahdavi, V.: All members of the MHC multigene family respond to thyroid hormone in a highly tissue-specific manner. Science *231:* 597–600 (1986).

39   Kaplan, J.H.; Forbush, B.; Hoffman, J.F.: Rapid photolytic release of adenosine 5′-triphosphate from a protected analogue: utilization by the Na:K pump of human red blood cell ghosts. Biochemistry *17:* 1929–1935 (1978).

40   Kress, M.; Huxley, H.E.; Faruqi A.R.; Hendrix, J.: Structural changes during activation of frog muscle studied by time-resolved X-ray diffraction. J. molec. Biol. *188:* 325–342 (1986).

41   Kushmerick, M.J.: Energetics of muscle contraction; in Peachey, Adrian, Geiger, Handbook of physiology, sect. 10: skeletal muscle, pp. 189–236 (American Physiological Society, Bethesda 1983).

42   Kushmerick, M.J.; Davies, R.E.: The chemical energetics of muscle contraction. II. The chemistry, efficiency and power of maximally working sartorius muscles. Proc. R. Soc. Lond. Ser. B *174:* 315–353 (1969).

43   Lowenstein, J.M.: Ammonia production in muscle and other tissues. The purine nucleotide cycle. Physiol. Rev. *52:* 382–414 (1972).

44   Lowey, S.: Cardiac and skeletal muscle myosin polymorphism. Med. Sci. Sports Exercise *18:* 284–291 (1986).

45   Lymn, R.W.; Taylor, E.W.: Mechanism of adenosine triphosphate hydrolysis by actomyosin. Biochemistry *10:* 4617–4624 (1971).

46   Margaria, R.: Biomechanics and energetics of muscular exercise, pp. 1–146 (Clarendon Press, Oxford 1976).

47   Martonosi, A.N.; Beeler, T.J.: Mechanism of Ca^{2+} transport by sarcoplasmic reticulum; in Peachey, Adrian, Geiger, Handbook of physiology, sect. 10: skeletal muscle, pp. 417–485 (American Physiological Society, Bethesda 1983).

48   Moss, R.L.; Allen, J.D.; Greaser, M.L.: Effects of partial extraction of troponin complex upon the tension-pCa relation in rabbit skeletal muscle. J. gen. Physiol. *87:* 761–774 (1986).

49   Moss, R.L.; Giulian, G.G.; Greaser, M.L.: Physiological effects accompanying the removal of myosin LC2 from skinned skeletal muscle fibers. J. biol. Chem. *257:* 8588–8591 (1982).

50   Natori, R.: Skinned fibres of skeletal muscle and the mechanism of muscle contraction. Jikeikai med. J. *33:* suppl. 1, pp. 1–74 (1986).

51   Rall, J.A.: Energetic aspects of skeletal muscle contraction: implications of fiber types; in Terjung, Exercise and sport sciences reviews, vol. 13, pp. 33–74 (MacMillan, New York 1985).

52   Reiser, P.M.; Moss, R.L.; Giulian, G.G.; Greaser, M.L.: Shortening velocity in single fibers from adult soleus muscles is correlated with myosin heavy chain composition. J. biol. Chem. *260:* 9077–9080 (1985).

53   Ruegg, J.C.: Calcium in muscle activation, pp. 1–300 (Springer, Berlin 1986).

54   Schneider, M.F.; Chandler, W.K.: Voltage dependent charge movement in skeletal muscle: a possible step in excitation-contraction coupling. Nature, Lond. *242:* 244–246 (1973).

55   Smith, I.C.H.: Energetics of activation in frog and toad muscle. J. Physiol. *220:* 583–599 (1972).

56   Somlyo, A.V.; McClellan, G.; Gonzalez-Serratos, H.; Somlyo, A.P.: Electron probe X-ray microanalysis of post-tetanic Ca^{2+} and Mg^{2+} movements across the sarcoplasmic reticulum in situ. J. biol. Chem. *260:* 6801–6807 (1985).

57   Stewart, M.; McLachlan, A.D.; Calladine, C.R.: A model to account for the elastic element in muscle crossbridges in terms of a bending myosin rod. Proc. R. Soc. Lond. B *229:* 381–413 (1987).

58   Terjung, R.L.; Dudley, G.A.; Meyer, R.A.: Metabolic and circulatory limitations to muscular performance at the organ level. J. exp. Biol. *115:* 307–318 (1985).

59   Wilkie, D.R.: The efficiency of muscular contraction. J. Mechanochem. Cell Motil. *2:* 257–267 (1974).

60   Wilkie, D.R.: Muscle; 2nd ed., pp. 1–68 (Edward Arnold, New York 1976).

61   Woledge, R.C.; Curtin, N.A.; Homsher, E.: Energetic aspects of muscle contraction, pp. 1–357 (Academic Press, London 1985).

Dr. Jack A. Rall, Department of Physiology, Ohio State University,
333 W 10th Avenue, Columbus, OH 43210 (USA)

Poortmans JR (ed): Principles of Exercise Biochemistry.
Med Sport Sci. Basel, Karger, 1988, vol 27, pp 22–39.

The Fibre Composition of Skeletal Muscle[1]

C.L. Rice, F.P. Pettigrew, E.G. Noble, A.W. Taylor

Faculty of Physical Education, The University of Western Ontario,
London, Ont., Canada

Introduction

The primary role of skeletal muscle is to generate force and this occurs
when motor units are activated. The motor unit was defined, at the beginning of the century, as the motor axon and the muscle fibres innervated by
it [49]. Contemporary researchers have included the soma of the nerve
fibre in this definition [9, 10]. Various techniques have shown that all the
muscle fibres within a single motor unit possess identical physiochemical
and morphological features in normal muscle [41]. However, muscle fibres
in different motor units may possess different physiological, biochemical
and morphological characteristics [10]. The purpose of this chapter is to
briefly review the basic concepts of fibre type classification and to discuss
the distribution and sizes of the different fibre types as they relate to
mammals, and specifically to humans.

The number of muscle fibres associated with one motoneuron is a
function of the size and role of the muscle. For example, the human vastus
lateralis, which is involved in the production of great force, may contain
motor units with more than 500 fibres per motoneuron, whereas the
adductor pollicis, which is involved in small and precise movements of the
hand, may contain motor units with only 5 or 6 muscle fibres per motoneuron. This variation in fibre number per motor unit is similar across all
mammalian species studied to date.

In nonhuman species, these findings have been determined using single motor unit electrical stimulation to deplete the glycogen stores from the

[1] Preparation of this manuscript was supported in part by a Gerontology Research
Council of Ontario Fellowship to F.P. Pettigrew, and Natural Sciences and Engineering Research Council (Canada) grants to E.G. Noble (No. A2897) and A.W. Taylor (No. A2787).

associated muscle fibres [9]. The muscle is then cut in serial cross-sections and treated histochemically to demonstrate the affected fibres related to the motor unit [63]. A different procedure for studying human muscles involves selective motor unit stimulation combined with electromyographic (EMG) and force measurement techniques [63]. In addition to innervation ratio differences among different motor units, these studies have also demonstrated that fibres from a particular unit are rarely adjacent and may occupy 20–33% of the mean cross-sectional area of the muscle, or 15% of its volume [9]. Any portion of a muscle is therefore composed of a mixture of different motor units.

A heterogeneous mixture of motor unit types is found in virtually all human muscles and most muscles of other mammals. The different characteristics of the muscle fibres representative of a motor unit have been extensively studied and categorized over the past several decades [20]. As a result, different fibre classification schemes have been developed based on contractile properties, metabolic markers, and histochemical and biochemical parameters.

Classification Schemes

There are many similarities between the various classifications, which are useful from a practical standpoint (table I). However, as has been recently emphasized [20, 24, 25], the schemes are not interchangeable, and as each reflects different characteristics of the muscle fibre, it is advisable

Table I. Fibre type classifications

Property	Fibre type				Reference No.
Contractile properties	S	FR-F (Int)		FF	9
Metabolic properties	SO	FOG	–	FG	55
Myosin ATPase activity	I	IIa	IIc	IIb	8
ATPase activity related to twitch properties	ST	FTa	–	FTb	60

SO = Slow, high oxidative activity; FOG = fast, high oxidative and glycolytic activities; FG = fast, high glycolytic activity; S = slow; FR = fast, fatigue resistant; F(Int) = fast, intermediate; FF = fast, fatiguable; ST = slow twitch; FT = fast twitch.

Table II. Contractile properties of different fibre types [10, 63] of non human mammals

Property	Fibre type			
	S	FR	F(Int)	FF
Twitch contraction time, ms	slow (58–110)	fast (30–55)	fast (33–40)	fast (20–47)
Maximum tetanic tension, g	low (2–13)	intermediate (5–55)	intermediate-high (13–40)	high (30–130)
Resistance to fatigue	very high	high	intermediate	low
'Sag' in unfused tetanus	absent	present	present	present
Post-tetanic potentiation	present (?)	present	present	present
Motor unit firing frequency	low (5–15 Hz)	intermediate (15–40 Hz)	intermediate	high (50–100 Hz)

Table III. Histochemical and biochemical properties of different fibre types [10, 17, 57, 60]

Property	Fibre type			
	I	IIa	IIc	IIb
Enzyme activity				
Myofibrillar ATPase	low	high	high	high
Oxidative enzymes	high	intermediate-high	intermediate	low
Glycolytic enzymes	low	high	high	high
Substrate concentration				
ATP	intermediate	high	intermediate	high
CP	intermediate	high	high	high
Glycogen[1]	low	intermediate	high	intermediate
Triglycerides	high	–	–	low

[1] In human skeletal muscle, glycogen concentrations are similar in all fibre types [17].

Table IV. Morphological characteristics of different fibre types [10, 16, 40, 56]

Property	Fibre type		
	ST	FTa	FTb
Mean fibre size, $\mu m^{2, a}$	1,730	2,890	5,290
Motor unit innervation ratio	540/U	440/U	750/U
Axonal conduction velocity, m/s	85	100	100
Fibre capillary supply	great	great	small
Mitochondria content	high	moderate	low
Sarcoplasmic reticulum (SR)	FT SR twice as extensive as ST SR		
Z band	ST Z band wider than FT Z band		
Connective tissue	collagen more abundant in ST than FT		

a Size is species dependent, this example is taken from cat muscle data.

to use the appropriate classification when discussing a particular fibre property. Further information about the techniques used in the derivation of the different classifications may be obtained from the references indicated in table I.

Generally in human studies, the type I and II, or ST and FT classifications are used since they directly reflect the common histochemical procedures employed (fig. 1) to assess the fibre characteristics [60]. Researchers employing contractile property measurements on animal models often use the Burke [9] classification, whereas metabolic characteristics measured through histo/biochemical procedures are commonly classified according to Peter et al. [55]. Tables II–IV provide summaries of some of the main fibre type characteristics associated with each methodology. For a more comprehensive understanding and review of these methods and their differences, the reader is referred to articles by Peter [56] and Burke and Edgerton [10].

Fibre Type Distribution

Although the information presented in tables II–IV represents the basic characteristics of the various fibre types, there are many important differences between the quality of these characteristics, as well as distribu-

Table V. Comparison of fibre type distribution of selected mammalian lower limb muscles [3, 6, 32, 60, 64]

Muscle	Human	Rat	Guinea pig	Cat	Dog	Horse
Lateral gastrocnemius						
% type I	50–64	28	12	18	50	–
% type II	40–52	72	88	82	50	–
Soleus						
% type I	70–90	96	100	100	–	–
% type II	10–30	4	–	–	–	–
Tibialis anterior						
% type I	55–65	2	4	19	32	–
% type II	35–45	98	96	81	68	–
Vastus lateralis						
% type I	40–50	9	–	27	43	30
% type II	50–60	91	100	73	57	70

tion patterns of the different fibre types between humans and other mammals. One must be cautious therefore when comparing a particular characteristic between species or between different muscles of the same individual [24]. It is also misleading to assume that the fibre type distribution profile from one muscle will be the same for all other muscles of the same individual [4, 11]. There are, however, accepted ranges of values for fibre type distribution for many mammalian limb muscles (table V). An excellent summary of the distribution profiles for many commonly studied human muscles has been provided by Saltin and Gollnick [60].

In humans, virtually all muscles consist of an heterogeneous mixture of fibre types (fig. 1), whereas there are several examples of muscles from other species (e.g. mouse soleus, guinea pig vasti) which are composed of mainly one fibre type (fig. 2). Within most species, the proportion of different fibre types for a particular muscle is relatively constant between individuals. Human skeletal muscle, on the other hand, demonstrates a large range of fibre distribution variability between individuals for a given muscle. For most human limb muscles, this variability represents a normal distribution throughout the population. In spite of the usual random heterogeneous mix of fibre types found in most human muscles, there are certain muscles or muscle groups that always present a very large propor-

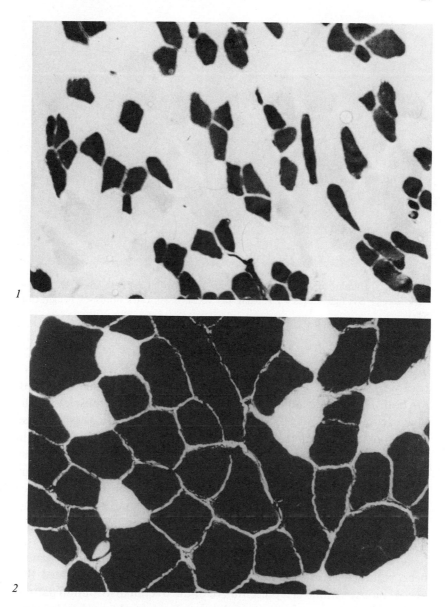

Fig. 1. Human gastrocnemius muscle (trained). Myofibrillar ATPase stain (pH 4.35). ×80. Dark fibres = ST; light fibres = FT.

Fig. 2. Horse gluteus medius muscle (trained). Myofibrillar ATPase stain (pH 10.2). ×159. Dark fibres = FT; light fibres = ST.

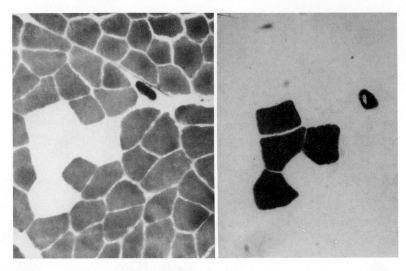

Fig. 3. Human soleus muscle (trained). Myofibrillar ATPase stain (left: pH 4.35, dark fibres = ST; light fibres = FT; right: pH 10.2). $\times$ 80. Dark fibres = FT; light fibres = ST.

tion of either ST fibres like the soleus (fig. 3), or FT fibres such as the triceps brachii [60]. These exceptions may have a functional basis in that, for both human and other animal species, antigravity muscles exhibit a higher composition of ST fibres compared with gravity assisted muscles which possess a greater proportion of FT fibres. There appears to be a regular stratification pattern of fibre types within certain limb muscles of lower mammals and, to some extent, in humans. The deepest portions of an antigravity muscle or muscle group usually possess a higher incidence of ST fibres [4]. The regional differences in fibre type distribution within a muscle probably reflect differences in motor unit location, and consequently activation.

The distribution of fibre types throughout human muscle has generally been regarded as random. For this reason, it has been assumed that a needle biopsy sample is representative of the overall distribution pattern of the muscle and, in some cases, of the muscle group. More detailed studies have been carried out over the past several years concerning the validity of this assumption [47, 52]. These studies suggest that, at best, within certain limb muscles, a large variability in fibre type proportion exists throughout an individual muscle. It has also been demonstrated that there are variations in fibre type occurrences within different portions of a fascicle [62]. Therefore, two to three needle biopsy samples have been proposed as the number required from a particular muscle to obtain an accurate estimate of the fibre type composition and a measure of the variability between samples.

Fibre Type Contractile Properties

The major function of muscle is to contract, and therefore to produce purposeful movement. It is logical to assume that the organization (distribution) of fibre types with different characteristics is a primary determinant of the various expressions of this function. It appears that under most circumstances there is an orderly recruitment pattern of different motor unit types [30]. In general, smaller motor units possess axons with the lowest firing threshold. These axons usually innervate ST fibres and the resulting forces are small but can be sustained for long periods of time. At the other end of the spectrum, large motor units possessing large high-threshold myelinated axons innervating FT fibres, are recruited for the production of great force and fire in brief high frequency bursts for a limited period of time. Between these two extremes, there are intermediate-sized motor units of different types, with intermediate properties. However, fast motor units (FF, F(Int), FR) are distinct from slow motor units (S) in all normal muscles (table II).

During muscle activity, a sharing effect between motor units of the same type and of different types exists. For contractions of low to moderate force levels both slow and fast fatigue-resistant, motor units are activated asynchronously. If the activity is carried out for an extended duration, those units (primarily slow units) which were preferentially recruited will begin to fatigue. As the muscle relies more and more on fast less-fatigue-resistant motor units, contractile force will decline. Synchronous firing of all the motor units in a muscle is a rare event resulting in a short duration contraction. Although the orderly recruitment of motor units (small to large) has been well established, it may be altered by pain stimuli, biofeedback training or rapid maximal voluntary contractions [14, 26, 31, 39].

As a general rule, larger animals exhibit slower contractile characteristics for all muscle fibre types [13]. For example, muscles composed primarily of ST fibres in the rat have faster contractile properties than muscles composed primarily of FT fibres in humans [13]. There is also evidence to indicate that ST fibres from one muscle possess different properties than ST fibres from another muscle within the same individual [58]. It is interesting to note that the relative proportions of fibre types are quite similar between homologous muscles of different species [10].

Fibre Type Distribution in Athletes

Cross-sectional comparisons of fibre type distribution from individuals actively involved in athletic activities have consistently demonstrated a selective dominance, depending on the sport, of either FT or ST fibres. Athletes engaged in endurance type events have shown a preponderance of ST fibres, whereas athletes engaged in less aerobic, nonendurance sports possess more FT fibres in selected limb muscles (table VII). A thorough review of these comparisons from a large number of sports can be found in Saltin and Gollnick's [60] review. There is, however, no conclusive evidence that success in an athletic event can be predicted on the basis of fibre type distribution [19]. In fact, considering the great variability in fibre type distribution between individuals, even in highly successful participants in the same athletic event, it is possible that participation and excellence in an event may be the result of several genetic predispositions.

Interconversion of ST and FT muscle fibre characteristics has been demonstrated in animals using cross-innervation techniques [13] and chronic nerve stimulation experiments [15]. It is clear from these experiments that the frequency, duration and coding of electrical stimuli from a motor neuron influence the basic fibre type characteristics of a muscle unit and many experiments have been conducted to determine whether various forms of exercise training in animals and humans can produce a shift in the fibre type profile of a particular muscle. Although it is well established that the metabolic and morphological profiles of a muscle can be altered with training [15, 60], the conversion of contractile characteristics and contractile proteins is, for the most part, inconclusive. Only a few longitudinal training studies have shown an alteration in the myofibrillar ATPase histochemical profile of exercised muscle [35, 36]. Similarly, a limited number of studies [23, 35] have reported changes in the contractile proteins with chronic exercise training. However, in a recent study, rats were subjected to an extreme endurance training regimen and subsequently demonstrated an increase in type I and IIa fibres, and a decrease in type IIb fibres

Table VI. Comparison of fibre sizes in lower limb muscles between rat, horse and human [5, 33, 60, 64]

Muscle	Species	Area, $\mu m^2 \times 10^2$	
		type I	type II
Lateral gastrocnemius	human	27–82	26–69
	rat	23	43
Soleus	human	48–90	71–147
	rat	41	25
Tibialis anterior	human	30–35	32–55
	rat	15	36
Vastus lateralis	human	30–92	22–94
	rat	26	48
	horse	30	33

Table VII. Fibre type distribution in trained and untrained males and females (cross-sectional comparisons) [38, 45, 60, 65, 67]

Muscle	Sex	% ST			% FT		
		un-trained	strength trained	endurance trained	un-trained	strength trained	endurance trained
Lateral gastrocnemius							
	M	50–64	21–27	63–74	40–52	70–80	25–35
	F	50–64	48–54	44–73	40–52	45–55	35–45
Vastus lateralis							
	M	40–50	25–60	53–70	50–60	25–65	35–45
	F	40–50	–	24–82	50–60	–	27–90
Deltoid							
	M	43–67	36–72	35–66	33–56	28–64	34–65
	F	–	–	–	–	–	–
Triceps brachii							
	M	14–46	19–61[a]	38–55	54–86	39–81[a]	45–62
	F	–	61[b]	–	–	39[b]	–

[a] Wheelchair athletes.
[b] One female wheelchair athlete.

[23]. Also, Howald [34] found evidence suggesting changes from type I fibres to fibres resembling type IIc fibres after strength training in humans.

Whether it is possible to produce a conversion between the two basic fibre types in response to an exercise stimulus has not been clearly demonstrated. To date, the evidence suggests that it may be possible, with endurance training, to alter the composition of the FT subtypes, that is, to increase IIa fibres and decrease IIb fibres [1, 15, 35]. It is clear that more rigorous longitudinal training studies are required before valid conclusions concerning fibre type transformation can be reached.

Fibre Area

The most striking difference between human and other animal fibre sizes is that type I fibres from animals in the untrained state are much smaller (table VI) than type II fibres (fig. 2), whereas in humans (fig. 3) ST and FT fibres are of nearly equal size [10, 60]. Although major alterations in fibre type distribution do not seem to occur with normal exercise training, changes in fibre size are demonstrable (table VIII).

Before discussing some of these changes, we will, for the purposes of this chapter, assume that the total number of fibres in any muscle is fixed at, or soon after, birth in both man and other mammals. Whether exercise training can cause changes in fibre number has been the subject of a great deal of study, but the evidence seems to indicate that hyperplasia of muscle fibres does not occur [18, 68]; only hypertrophy and atrophy [60, 66]. For alternative views of this controversy, see recent papers by Gollnick et al. [18], Gonyea et al. [21], Larsson and Tesch [44], and Timson [68].

Endurance training in animals has for the most part resulted in decreases in muscle fibre size. Studies in humans, however, have given divergent results. The data indicate either no change, selective increase, or decrease in ST fibre size with endurance training [15]. Cross-sectional comparison, biopsy sample number, and lack of standardization with respect to training intensity, program duration and specificity of the sampled muscle are major problems which may account for the discrepancies in fibre adaptation seen in human studies.

Resistance or strength training has been observed in both animals and humans to produce fibre enlargement. The macroscopic result is an overall enlargement or volume increase of the muscle group due to the sum total

Table VIII. Fibre type area measurements in trained and untrained males and females (cross-sectional comparisons) [38, 45, 60, 65, 67]

Muscle	Sex	ST area, $\mu m^2 \times 10^2$			FT area, $\mu m^2 \times 10^2$		
		un-trained	strength trained	endurance trained	un-trained	strength trained	endurance trained
Lateral gastrocnemius							
	M	45–82	40–78	36–101	32–51	59–62	47–113
	F	27–72	33–71	31–100	25–50	48–69	40–75
Vastus lateralis							
	M	23–42	17–101	34–69	22–52	58–145	25–69
	F	20–36	–	31–72	17–42	–	15–77
Deltoid							
	M	27–67	38–66	26–65	40–80	65–106	36–76
	F	–	–	–	–	–	–
Triceps brachii							
	M	67–80	63–125[a]	72–92	104–133	93–187[a]	125–163
	F	–	120[b]	–	–	146[b]	–

[a] Wheelchair athletes.
[b] One female wheelchair athlete.

enlargement of the individual fibres. As indicated earlier, the overall enlargement is presently not considered a result of longitudinal fibre splitting or hyperplasia. The increased strength associated with fibre hypertrophy is believed to result mainly from a quantitative increase in the myofibrillar protein content and not qualitative changes [60]. These quantitative changes may result from longitudinal splitting of existing myofibrils [53].

In mixed muscles of both humans and other animals, weight training causes a greater enlargement of FT fibres. In fact, it has been reported that body builders possess extremely large FT fibres [51]. In other animals homogeneous slow twitch muscles show smaller fibre size increases than homogeneous fast twitch muscles [15]. This selective adaptation may be related to the training stimulus (maximal efforts for relatively short duration) which recruits FT motor units to a greater extent than during normal physical activity [53].

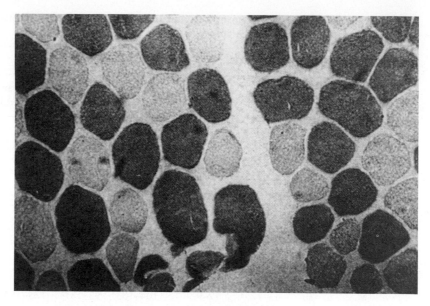

Fig. 4. Human vastus lateralis muscle (postcoronary infarct patient, bedridden for 3 months). Myofibrillar ATPase stain (pH 10.0). $\times$ 159. Dark fibres = FT; light fibres = ST.

It is well recognized that training-induced fibre alteration in animal muscle leads to changes of a greater magnitude than in humans. This does not necessarily imply that animal muscle is more mutable, but more likely suggests that the important changes observed are a result of very intense training for prolonged periods of time. It is questionable whether human subjects could endure relatively similar training regimens to produce analogous alterations [53].

Detraining and to some degree immobilization of a muscle (fig. 4) both affect the muscle fibres similarly. Any fibre change induced by exercise training will return to the pre-exercise state after cessation of the training stimulus. Whether the time course of the decline is the same as the training period has not been conclusively established [43, 50, 61]. Continued immobilization will result in fibre atrophy and possibly a change in the fibre distribution profile [34, 60]. It has been suggested that the reduction in fibre size will preferentially affect ST fibres, FT fibres, or both. A review of this controversy has recently been published by Appell [2].

Fibre Type Distribution and Ageing

Advanced age is characterized by muscle atrophy and concomitant loss of muscle strength. Most researchers agree that much of the loss in force generating capacity is due to a decline in the quantity of muscle tissue. Cross-sectional studies comparing young and old individuals have demonstrated dramatic reductions (25–50%) in overall fibre number as determined histochemically at autopsy [46], using the muscle biopsy technique [48], and electrophysiologically [22, 27]. Loss of muscle fibres with age has also been demonstrated in other mammals [7, 12, 59]. It has been suggested that the loss of muscle fibres is either secondary to a loss of motoneurons or a result of degeneration of the muscle fibres per se [29, 42].

The number of muscle fibres in man is believed to continuously decline from a maximum number at birth. The functional significance of this loss, however, is not evident until 60–70 years of age. Other mammalian species also demonstrate a more pronounced loss of muscle mass and functional decline at the same relative point in their lifespan [27]. Although earlier studies pointed to a preferential loss of FT fibres in ageing skeletal muscle [42], these findings have not been substantiated and the current consensus is that the age-related loss of fibres is not type specific [22, 28].

In normal individuals younger than 70 years of age, most skeletal muscles demonstrate a mosaic pattern of fibre type distribution with only occasional small groups of homogeneous fibres. Beyond 70 years of age, however, the mosaic pattern in many muscles is replaced with groups of fibres of one type [54]. This fibre type grouping is expressed in clusters of 50 or more ST fibres [37] and probably occurs as a result of FT fibres becoming reinnervated by axons innervating surviving ST muscle units [46]. In general, distal muscles are more severely affected by fibre type grouping, suggesting that part of the process may be due to disruption of the peripheral axons. Eventually, the process of denervation-reinnervation reaches a point where all the fibres become permanently denervated, degenerate and are replaced by adipose and connective tissue [48].

Fibre size alterations as a result of ageing are not dramatic and, therefore, most of the loss of muscle mass is primarily attributable to a loss of muscle fibres. Slight reductions in FT fibre size have been found in some studies, but the differences are not significant [27]. Overall, women possess smaller muscle masses than men, but with ageing, it has been noted that the reduction in muscle mass is more pronounced in men.

Information as to the effect of exercise on aged skeletal muscle is very limited, but it has been suggested that the potential for muscle adaptability exists [27]. The extent to which skeletal muscle of aged individuals is capable of positive adaptations is still to be determined. It has been proposed that neural factors play a greater role in the trainability of old muscles as compared to young muscles. Review papers by Green [22] and Grimby and Saltin [27] summarize the available information relating the effects of exercise and ageing on human skeletal muscle.

References

1   Andersen, P.; Henriksson, J.: Training induced changes in the subgroups of human type II skeletal muscle fibers. Acta physiol. scand. 99: 123–125 (1977).
2   Appell, H.-J.: Skeletal muscle atrophy during immobilization. Int. J. Sports Med. 7: 1–5 (1986).
3   Ariano, M.A.; Armstrong, R.B.; Edgerton, V.R.: Hindlimb muscle fibre population of five mammals. J. Histochem. Cytochem. 21: 51–55 (1973).
4   Armstrong, R.B.: Properties, distribution and function of mammalian skeletal muscle fibres; in Cerretelli, Whipp, Exercise bioenergetics and gas exchange, pp. 137–146 (Elsevier/North Holland/Biomedical Press, Amsterdam 1980).
5   Armstrong, R.B.; Phelps, R.O.: Muscle fibre type composition of the rat hindlimb. Am. J. Anat. 171: 259–272 (1984).
6   Armstrong, R.B.; Saubert, C.W., IV; Seehernon, H.J.; Taylor, C.R.: Distribution of fibre types in locomotory muscles of dogs. Am. J. Anat. 163: 87–98 (1982).
7   Bass, A.; Gutman, E.; Hanzlikova, V.: Biochemical and histochemical changes in energy-supply-enzyme pattern of muscles of rat during old age. Gerontologia 21: 31–45 (1975).
8   Brooke, M.H.; Kaiser, K.K.: Muscle fibre types. How many and what kind? Archs Neurol. 23: 369–379 (1970).
9   Burke, R.E.: Motor units: anatomy, physiology and functional organization; in Brooks, Handbook of physiology, pp. 345–422 (American Physiological Society, Bethesda 1981).
10 Burke, R.E.; Edgerton, V.R.: Motor unit properties and selective involvement in movement. Exercise Sports Sci. Rev. 1975: 31–81.
11 Burke, R.E.; Tsairis, P.: The correlation of physiological properties with histochemical characteristics in single muscle units. Ann. N.Y. Acad. Sci. 228: 145–159 (1974).
12 Caccia, M.C.; Harris, J.B.; Johnson, M.A.: Morphology and physiology of skeletal muscle in aging rodents. Muscle Nerve 2: 202–212 (1979).
13 Close, R.I.: Dynamic properties of mammalian skeletal muscles. Physiol. Rev. 52: 129–197 (1972).

14 Desmedt, J.E.: Patterns of motor commands during various types of voluntary movements in man. Trends Neurosci. *3:* 265–268 (1980).

15 Edstrom, L.; Grimby, L.: Effect of exercise on the motor units. Muscle Nerve *9:* 104–126 (1986).

16 Eisenberg, B.R.: Adaptability of ultrastructure in the mammalian muscle. J. exp. Biol. *115:* 55–68 (1985).

17 Essen, B.; Jansson, E.; Henriksson, J.; Taylor, A.W.; Saltin, B.: Metabolic characteristics of fibre types in human skeletal muscle. Acta physiol. scand. *95:* 153–165 (1975).

18 Gollnick, P.D.; Timson, B.F.; Moore, R.L.; Riedy, M.: Muscular enlargement and number of fibers in skeletal muscle of rats. J. appl. Physiol. *50:* 936–943 (1981).

19 Gollnick, P.D.; Matoba, H.: The muscle fibre composition of muscle as a predictor of athletic success. Am. J. Sports Med. *12:* 212–217 (1984).

20 Gollnick, P.D.; Hodgson, D.R.: The identification of fibre types in skeletal muscle: a continual dilemma. Exercise Sports Sci. Rev. *1986:* 81–104.

21 Gonyea, W.J.; Sale, D.G.; Gonyea, F.B.; Milkesky, A.: Exercise induced increases in muscle fibre number. Eur. J. appl. Physiol. *55:* 137–141 (1986).

22 Green, H.J.: Characteristics of aging human skeletal muscles; in Sutton, Brock, Sports medicine for the mature athlete, pp. 17–26 (Benchmark Press, Columbia 1986).

23 Green, H.J.; Klug, G.A.; Reichmann, H.; Seedorf, U.; Wrieher, W.; Pette, D.: Exercise induced fibre type transitions with regard to myosin, parvalbumin and sarcoplasmic reticulum in muscles of the rat. Pflügers Arch. *400:* 432–438 (1984).

24 Green, H.J.; Reichmann, H.; Pette, D.: Inter- and intraspecies comparisons of fibre type distribution and of succinate dehydrogenase activity in type I, IIa and IIb fibres of mammalian diaphragms. Histochemistry *81:* 67–73 (1984).

25 Green, H.; Reichmann, H.; Pette, D.: A comparison of two ATPase based schemes for histochemical muscle fibre typing in various mammals. Histochemistry *76:* 21–31 (1982).

26 Grimby, L.; Hannerz, J.: Flexibility of recruitment order of continuously and intermittently discharging motor units in voluntary contraction; in Desmedt, Motor unit types, recruitment and plasticity in health and disease. Prog. clin. Neurophysiol., vol. 9, pp. 201–211 (Karger, Basel 1981).

27 Grimby, G.; Saltin, B.: The ageing muscle. Mini-review. Clin. Physiol. *3:* 209–218 (1983).

28 Grimby, G.; Ariansson, A.; Zetterberg, C.; Saltin, B.: Is there a change in muscle fibre composition with age? Clin. Physiol. *4:* 189–194 (1984).

29 Gutmann, E.; Hanzlikova, V.: Fast and slow motor units in ageing. Gerontology *22:* 280–300 (1976).

30 Henneman, E.; Mendell, L.M.: Functional organization of motoneurone pool and its inputs; in Brooks, Handbook of physiology, pp. 423–507 (American Physiological Society, Bethesda 1981).

31 Henneman, E.; Shahani, B.T.; Young, R.R.: Voluntary control of human motor units; in Shahani, Bombay Symposium on Motor Control, pp. 73–78 (Elsevier, Amsterdam 1976).

32 Henriksson-Larsen, K.B.; Lexell, J.; Sjostrom, M.: Distribution of different fibre

types in human skeletal muscles. I. Methods for the preparation and analysis of cross-sections of whole tibialis anterior. Histochem. J. *15:* 167–178 (1983).

33   Henriksson-Larsen, K.B.; Friden, J.; Wretling, M.-L.: Distribution of fibre sizes in human skeletal muscle. An enzyme histochemical study in m. tibialis anterior. Acta physiol. scand. *123:* 224–235 (1985).

34   Howald, H.: Malleability of the motor system: training for maximizing power output. J. exp. Biol. *115:* 365–373 (1985).

35   Ingjer, F.: Effects of endurance training on muscle fibre ATPase activity, capillary supply and mitochondrial content in man. J. Physiol. *294:* 419–432 (1979).

36   Jansson, E.; Kaijser, L.: Muscle adaptation to extreme endurance training in man. Acta physiol. scand. *100:* 315–324 (1977).

37   Jennekens, F.G.I.; Tomlinson, B.E.; Walton, J.N.: Histochemical aspects of five limb muscles in old age. J. neurol. Sci. *14:* 259–276 (1971).

38   Johnson, M.A.; Polgar, S.; Weightman, D.; Appleton, D.: Data on the distribution of fibre types in thirty-six human muscles. J. neurol. Sci. *18:* 111–129 (1973).

39   Kanda, K.; Burke, R.E.; Walmsley, B.: Differential control of fast and slow twitch motor units in the decerebrate cat. Exp. Brain Res. *29:* 57–74 (1977).

40   Kovanen, V.; Suominen, H.; Heikkinen, H.: Mechanical properties of fast and slow skeletal muscle with special reference to collagen and endurance training. J. Biochem. *17:* 725–735 (1984).

41   Kugelberg, E.: Adaptive transformation of rat soleus motor units during growth. J. neurol. Sci. *27:* 269–289 (1976).

42   Larsson, L.: Aging in mammalian skeletal muscle; in Mortimer, Pirozzolo, Maletta, The aging motor system, pp. 60–97 (Praeger, New York 1982).

43   Larsson, L.; Ansved, T.: Effect of long-term physical training and detraining on enzyme histochemical and functional skeletal muscle characteristics in man. Muscle Nerve *8:* 714–722 (1985).

44   Larsson, L.; Tesch, P.A.: Motor unit fibre density in extremely hypertrophied skeletal muscles in man. Electrophysiological signs of muscle fibre hyperplasia. Eur. J. appl. Physiol. *55:* 130–136 (1986).

45   Lavoie, J.-M.; Taylor, A.W.; Montpetit, R.R.: Skeletal muscle fibre size adaptation to an eight-week swimming programme. Eur. J. appl. Physiol. *44:* 161–165 (1980).

46   Lexell, J.; Henriksson-Larsen, K.; Winbald, B.; Sjostrom, M.: Distribution of different fibre types in human skeletal muscles. Effects of aging studied in whole muscle cross-sections. Muscle Nerve 6: 588–595 (1983).

47   Lexell, J.; Taylor, C.; Sjostrom, M.: Analysis of sampling errors in biopsy techniques using data from whole muscle cross-sections. J. appl. Physiol. *59:* 1228–1235 (1985).

48   Lexell, J.; Downham, D.; Sjostrom, M.: Distribution of different fibre types in human skeletal muscles. J. neurol. Sci. *72:* 211–222 (1986).

49   Liddell, E.G.T.; Sherrington, C.S.: Recruitment and some other factors of reflex inhibition. Proc. R. Soc. Lond. *B97:* 488–518 (1925).

50   Lindboe, C.F.; Platou, C.S.: Effect of immobilization of short duration on muscle fibre size. Clin. Physiol. *4:* 183–188 (1984).

51   MacDougall, J.D.; Sale, D.G.; Alway, S.E.; Sutton, J.R.: Muscle fibre number in biceps brachii in body builders and control subjects. J. appl. Physiol. *57:* 1399–1403 (1984).

52 Mahon, M.; Toman, A.; William, P.L.T.; Bagnall, K.M.: Variability of histochemical and morphometric data from needle biopsy specimens of human quadriceps femoris muscle. J. neurol. Sci. 63: 85–100 (1984).

53 McDonagh, M.J.N.; Davies, C.T.M.: Adaptive response of mammalian skeletal muscle to exercise with high loads. Eur. J. appl. Physiol. 52: 139–155 (1984).

54 Oertel, G.: Changes in human skeletal muscle due to ageing. Acta neuropath. 69: 309–313 (1986).

55 Peter, J.B.; Barnard, R.J.; Edgerton, V.R.; Gillespie, C.A.; Stempel, K.E.: Metabolic profiles of three fibre types of skeletal muscle in guinea pigs and rabbits. Biochemistry 11: 2627–2633 (1972).

56 Peter, J.B.: Skeletal muscle: diversity and mutability of its histochemical, electron microscopic, biochemical and physiological properties; in Pearson, Mostofi, The striated muscle, pp. 1–18 (Williams & Wilkins, Baltimore 1973).

57 Pette, D.: Metabolic heterogeneity of muscle fibres. J. exp. Biol. 115: 179–189 (1985).

58 Round, J.M.; Jones, P.A.; Chapman, S.J.; Edwards, R.H.T.; Ward, P.S.; Fodden, D.L.: The anatomy of fibre type composition of the human adductor pollicis in relation to its contractile properties. J. neurol. Sci. 66: 263–292 (1984).

59 Rowe, R.W.D.: The effect of senility on skeletal muscles in the mouse. Exp. Gerontol. 4: 119–126 (1969).

60 Saltin, B.; Gollnick, P.D.: Skeletal muscle adaptability: significance for metabolism and performance; in Peachey, Adrian, Geiger, Handbook of physiology, sect. 10: skeletal muscle, pp. 555–663 (American Physiological Society, Williams & Wilkins, Bethesda/Baltimore 1983).

61 Schantz, P.; Henriksson, J.; Jansson, E.: Training-induced increase in myofibrillar ATPase intermediate fibers in human skeletal muscle. Clin. Physiol. 3: 141–151 (1983).

62 Sojstrom, M.; Downham, D.Y.; Lexell, J.: Distribution of different fibre types in human skeletal muscles. Why is there a difference within a fascicle? Muscle Nerve 9: 30–36 (1986).

63 Stuart, D.G.; Enoka, R.M.: Motoneurons, motor units and the size principle; in Rosenberg, The clinical neurosciences, vol. 17, pp. 471–517 (Churchill Livingstone, New York 1983).

64 Taylor, A.W.; Brassard, L.: Skeletal muscle fibre distribution and area in trained and stalled standardbred horses. Can. J. anim. Sci. 61: 601–605 (1981).

65 Taylor, A.W.; McDonnell, E.; Royer, D.; Loiselle, R.; Lush, N.; Steadward, R.: Skeletal muscle analysis of wheelchair athletes. Paraplegia 17: 456–460 (1979).

66 Taylor, N.A.S.; Wilkinson, J.G.: Exercise-induced skeletal muscle growth, hypertrophy or hyperplasia? Sports Med. 3: 190–200 (1986).

67 Tesch, P.A.; Karlsson, J.: Muscle fibre types and size in trained and untrained muscles of elite athletes. J. appl. Physiol. 59: 1716–1720 (1985).

68 Timson, B.F.: The effect of varying postnatal growth rate on skeletal muscle fiber number in the mouse. Growth 46: 36–45 (1982).

Dr. A.W. Taylor, Faculty of Physical Education, The University of Western Ontario, London, Ont. N6A 3K7 (Canada)

Poortmans JR (ed): Principles of Exercise Biochemistry.
Med Sport Sci. Basel, Karger, 1988, vol 27, pp 40–77.

Basic Aspects of Metabolic Regulation and Their Application to Provision of Energy in Exercise

E.A. Newsholme

Department of Biochemistry, University of Oxford, UK

Introduction

Energy expenditure is increased when the body carries out physical work. The efficiency with which this energy is expended in relation to the work done has been investigated in depth and it is usually considered to be about 25% [13]. The implicit assumption is that all of the energy not transformed into work can be classified as wasted. However, this must represent a myopic view of energy consumption since it does not take into account the energy that is expended to provide for control and regulation of all the processes involved in the expenditure of energy. Regulation occurs at many different levels and the preoccupation of biologists and physiologists with nervous and endocrine control may explain a prevalent viewpoint; that the amount of energy expended in regulation is quantitatively trivial. A recent quantitative approach into the subject of metabolic regulation provides at least the basis not only for testing model systems of regulation but also for investigating the energy cost of metabolic regulation [11, 14]. The magnitude of the problem that faces the body is illustrated by consideration of just three everyday situations.

(1) A normal western diet contains a high proportion of carbohydrate; approximately 300 g of carbohydrate is ingested everyday. If this is consumed equally during three meals a day this means that more than 100 g are consumed at each meal, which is sufficient to increase the blood glucose level more than 30-fold. That this does not happen is due to the precision of the mechanisms that control the blood glucose level. Some of

the extra glucose absorbed will be converted to glycogen in muscle and liver which have the capacity to store over 400 g of glucose [16]. However, the *total* capacity for the rate of synthesis of glycogen in man is probably more than 1 g of glucose per min so that if feeding caused complete activation of glycogen synthase, it could rapidly result in hypoglycaemia, coma and possibly death. Consequently, eating is a dangerous activity; it is made safe by the precision of control mechanisms!

(2) In order to complete the 100-metre sprint in a respectable time the rate of conversion of glycogen to lactate must be increased more than 1,000-fold: in other words, from resting to maximum acceleration, the activities of all the enzymes of glycolysis must increase by more than 1,000-fold [16]. Similarly, for the average individual walking rapidly upstairs or running to catch the last bus or train probably requires an increase in glycolysis of several-hundred-fold. However, the control must be precise not only for increasing the rate of anaerobic glycolysis to a sufficient level to provide the ATP for the physical activity, but also for decreasing it! It can be calculated that capacity for maximum glycolysis in human muscle is such that if the glycolytic enzymes were fully activated, enough protons would be produced in about 4 min to lower the pH of the body to 2.0! Consequently, walking upstairs is a dangerous activity; it is made safe by the precision of control mechanisms!

(3) During endurance exercise (e.g. marathon running) the fat store in the adipose tissue is made available in the bloodstream as fatty acid. The hydrolysis of triacylglycerol occurs in the adipose tissue and fatty acids are transported to the muscle via the bloodstream. This provides an extremely important source of energy for muscle during endurance exercise [16]. It can be calculated that, if the rate of mobilisation of fatty acid was over-stimulated by just 1%, in about 60 min the concentration of plasma fatty acid could double to reach a concentration of perhaps 4 mM, which would be highly dangerous. Endurance exercise is a dangerous activity; it is made safe by the precision of control mechanisms!

In order to understand the overall approaches to metabolic control and, importantly, what the authors consider should be called *'metabolic control logic',* it is necessary to appreciate the theory underlying metabolic control.

The authors consider that the application of metabolic-control-logic to problems and questions relating to exercise physiology, exercise biochemistry and even exercise pathology has in the past and will, particularly in the future, provide new insights and new concepts.

Theoretical Basis to Understanding Metabolic Control

The Concept of a Metabolic Flux

Discussions of metabolism, its integration and control and questions that arise in relation to such problems as fuel supply for exercise, are usually based on the conventional biochemical pathways. These are given the stamp of approval by their existence as a separate chapter in textbooks or as part of a large and complex metabolic wall chart. Such pathways have been isolated by the biochemist as a series of enzyme-catalysed reactions that carry out a certain function in a cell. Consideration of the importance of steady state flux through a metabolic pathway and the control of this flux has led to the development of the concept of a physiological pathway or a metabolic flux, which extends the biochemical interpretation of a metabolic pathway [15]. The deficiency of the conventional biochemical pathways is that they are not always able, by themselves, to generate or maintain a steady state flux. As an example, consider the pathway of glycolysis-from-glucose in muscle. It is well established that the flux in this pathway can be increased by raising the blood glucose concentration above the physiological level of about 5 mM. That is, there is no reaction in this conventional pathway that is saturated with pathway substrate. Hence, utilisation of glucose by muscle could rapidly lead to a reduction in the blood glucose level which, in turn, would reduce the rate of glycolysis. In other words, the conventional glycolytic pathway is incomplete; there must be some means of maintaining a constant flux through glycolysis (if only for a limited time) by release of glucose into the bloodstream at a rate equal to the rate of utilisation by the muscles. This is achieved either by glucose absorption from the intestine (in the absorptive state) or by glucose release from the liver (in the post-absorptive state). These processes can maintain a constant rate since they contain a flux-generating step. The latter is a reaction that approaches saturation with pathway substrate, so that any decrease in the concentration of this substrate will not significantly reduce the rate of the reaction [15]. For the absorption of glucose from the intestine, the flux-generating step is probably the transport across the luminal membrane of the epithelial cells; for glucose release from glycogen in the liver, the flux-generating step is glycogen phosphorylase (fig. 1).

In the post-absorptive state, metabolic flux through glycolysis is initiated at the reaction catalysed by hepatic phosphorylase and this flux is then transmitted via the concentration of glucose in the bloodstream to muscle where it can be considered to end in the formation of pyruvate.

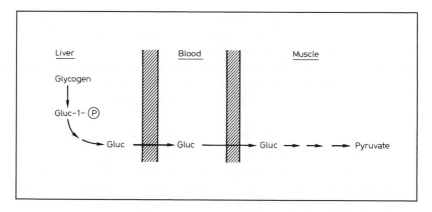

Fig. 1. Flux-generating steps for muscle glycolysis. During prolonged (e.g. marathon race) aerobic exercise in the post-absorptive state, glucose is converted to pyruvate via glycolysis in the muscle and pyruvate is oxidised by the Krebs cycle. This will utilise some blood glucose but the blood glucose concentration remains constant during races even as long as the marathon. The blood glucose is maintained by the breakdown of *liver* glycogen, which is released into the blood as glucose. The flux-generating step for this process is *glycogen phosphorylase* which catalyses the conversion of glycogen to glucose-6-phosphate in the liver.

Hence, the pathway spans more than one tissue. It will, of course, be realised that this represents an oversimplification of the situation in vivo since the flux generated by hepatic phosphorylase in the liver is divided between several tissues (e.g. kidney, brain, intestine as well as many different muscles) and the rate of utilisation will vary between tissues and between one muscle and another. The use of glucose by the different tissues produces a branched flux. The importance of branching in such a physiological pathway will be discussed below.

Control of the Transmission of a Flux
In the following hypothetical linear pathway the flux (J) is generated at reaction E_1 (the flux-generating step which is denoted by the sign $-\!/\!/\!\rightarrow$).

$$S \;-\!/\!/\!\!\rightarrow\; A \xrightarrow{\overset{X}{\overset{\oplus}{}}} B \longrightarrow C \longrightarrow (J)$$

$$\quad E_1 \qquad E_2 \quad\; E_3 \quad\; E_4$$

This flux is transmitted along the pathway by the response of the subsequent reactions to the metabolic intermediates A, B and C, so that these metabolites link all the component reactions of the pathway and help to produce the overall steady state flux. The role of the metabolic intermediates can be best explained by reference to the effect of an activation of one of the component reactions. Compound X stimulates enzyme E_2 so that, if the concentration of X was increased, the activity of this enzyme would increase. However, this would result in a decrease in the concentration of the substrate A which would lower the activity of enzyme E_2. The decrease in A will continue until the activity of E_2 reaches the original activity, so that the overall steady state flux will not change. The only difference would be a lowered concentration of A, so that the increase in activity of E_2 is only transient. This shows that, in this example, the concentration of A is determined by the flux (i.e. the activity of the enzyme that catalyses the flux-generating step, E_1) together with the kinetic properties of enzyme E_2. Similarly, the concentrations of B and C are determined by the flux and the kinetic properties of E_3 and E_4, respectively. Metabolic intermediates whose concentrations are determined by the flux (and which, therefore, help to maintain the steady state flux) are termed *internal* effectors for that flux. From this definition, it should be clear that internal effectors *cannot* change the flux. (They can be regarded as being totally enclosed by the flux.)

Regulators and Regulatory Reactions. The term regulation can be described and defined in terms of metabolic communication [15] which is another way of describing a stimulus-response system. The statement that P communicates with Q is another way of saying that a change in P produces a change in Q (this relationship is abbreviated as follows P ⤳ Q). Therefore, P can be considered as a regulator of Q since it is one factor that determines the value of Q. Consequently, any effector of an enzyme can be termed a regulator of that enzyme's activity. Of considerable importance in the interpretation of the properties of an enzyme in vitro is that effectors of the enzyme will not necessarily regulate the rate of that reaction in vivo. For example, in the system given above, both A and X are regulators of the enzyme E_2 (since they both communicate with the enzyme) but, as indicated above, X cannot change the rate of E_2 in vivo in the steady state (any change in the rate of E_2 is compensated for by the inverse change in the concentration of A). On the other hand, the internal effector A balances the rate of E_2 to the flux and hence is a regulator (an internal effector or regulator) of the rate of E_2.

If the hypothetical system given above is modified so that Y is a regulator of E_1, a comparison of the two effects is useful.

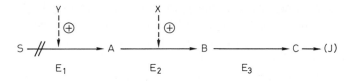

In contrast to X, regulator Y is able to change the rate of E_1 and hence the flux through the pathway. The important difference is that E_1 is a flux-generating step, so that any change in the concentration of S, in response to a change in Y, will not modify the activity of E_1, and the latter communicates with all the reactions that transmit the flux, via the concentrations of the pathway substrates A, B and C. Note that E_2 can communicate with E_3 via B but is unable to communicate with the flux-generating step E_1. However, if the system is modified to include an inhibitory effect of A on enzyme E_1, it is as follows:

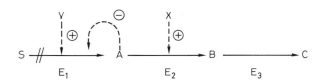

This establishes a feedback communication from E_2 to E_1 so that both E_1 and E_2 communicate with *all* the reactions that constitute the flux. Reactions that communicate with the flux are defined as *regulatory* for the flux. Although this may appear to be very straightforward, some care is needed when defining the flux that is being modified, especially when the pathway is branched (see below).

It has been emphasised that, in the absence of the feedback communication between E_2 and E_1, compound X cannot communicate with the flux, and, therefore, cannot regulate the activity of E_2. However, such an effect may not be redundant in vivo. An increase in the rate of E_1, produced by a change in concentration of Y is communicated to E_2 via the increase in concentration of A. However, if at the same time as the concentration of Y was increased, that of X was also increased, the change in the concentration of A needed to communicate between E_1 and E_2 would be less. This would be important if the concentration of A was utilised in some other pathway (i.e. if the pathway was branched), if A was a labile or a chemically active intermediate so that a high concentration would lead to unwanted side reactions, or if the change in flux had to be very large and the concentration change in A would not be able to produce the necessary increase in rate of E_2. Such considerations may explain why many enzymes in the physiological pathway, although they do not catalyse flux-generating steps, are subject to regulation by factors other than the pathway substrate (i.e. other than internal regulation). Effectors other than the pathway substrate are known as external effectors [15] and they can include allosteric effectors as well as cofactors or cosubstrates.

The above discussion emphasises that, in order to investigate the mechanism of regulation of a pathway, the complete physiological pathway should be known. For example, the precise role of external regulators, such as X in the above system, can only be fully ascertained in relation to a knowledge of the *complete* physiological pathway. If the conventional (i.e. textbook) pathway started at A, whereas the physiological pathway actually started at S, the effects of X on the control of the flux through the pathway could be easily misinterpreted. For the conventional pathway, it could be concluded that X was an important regulator of flux whereas, in fact, it is only a regulator of the concentration of A and *not* the flux. This is not merely an academic point. In biochemical experiments it is often very much easier to study conventional or truncated pathways rather than the complete physiological pathway and, for this, the pathway substrate (i.e. A) is present either in large amounts (to produce a pseudo-flux-generating step [15]) or at a concentration that approaches saturation with the enzyme (i.e. at an unphysiologically high concentration). Under these experimental conditions, changes in X would change the flux through the pathway giving rise to a false mechanism for the regulation of the flux through the pathway in vivo.

Regulatory Reactions. Although we have used the term regulatory effector to mean a compound that can affect the activity of an enzyme but not necessarily the flux, regulatory reactions are restricted to those that can communicate with the flux, J. Such reactions are termed regulatory for the flux [15]. In a similar manner to internal and external effectors, it is necessary to relate the term regulatory to a defined flux, since a regulatory reaction for one flux may not be regulatory for another (for example, in a branched pathway [3]).

From the above discussions it might seem that the flux-generating step in a physiological pathway must be the 'rate-limiting' or 'pacemaker' step for the flux. Although in some pathways this may be the case, if there are several regulatory reactions (and this is usually the case) none of these can be regarded as 'rate limiting'. A well-known example of a pathway with several regulatory steps is the conversion of glycogen to lactate in muscle; both 6-phosphofructokinase and phosphorylase are regulatory for this flux [16].

An important theoretical and practical point to emerge from these considerations is in relation to the reactions that will be modified by hormones. If a hormone is to change the flux through a metabolic pathway it must modify the activity of a regulatory reaction (i.e. either the flux-generating step or a reaction that can communicate with it). Since hepatic glycogen phosphorylase is a flux-generating reaction for glycolysis in muscle (and indeed other tissues) it is not surprising that its activity is modified by a variety of hormones including glucagon, catecholamines, and vasopressin. Similarly, glycogen synthetase in muscle and liver, triglyceride lipase in adipose tissue catalyse flux-generating steps, and, consequently, are influenced by a variety of hormones.

A further important point is that these theoretical principles provide information as to which enzymes would *not* be suitable targets for hormone action. That is, any reaction which is not regulatory or flux generating will not be influenced by a hormone to change the flux through that given physiological pathway. Although this may seen very obvious, there are many examples of such reactions that appear to be sensitive to hormones in vitro and this property is investigated in vitro in detail. For example, it is still reported in the many texts that steroid hormones modify hepatic glutamate dehydrogenase which, in the rat, is considered to catalyse a near-equilibrium reaction. Since this reaction is not regulatory for amino acid metabolism, the hormone effect is considered to be of no importance in vivo.

The long-established stimulatory effect of insulin on glucose transport in muscle can only be of physiological importance in relation to entry of glucose during the absorptive phase of digestion. A stimulatory effect of insulin in the absence of glucose absorption would produce hypoglycaemia and coma, since it is stimulating a non-flux-generating step so that the steady state flux would not be changed, but the concentration of the intermediate blood glucose will fall. This explains the fact that insulin can be used as a murder weapon!

Regulatory Sequences. In the following system

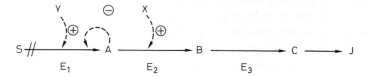

in which E_1 and E_2 are both regulatory for the flux (J), the effector Y communicates with the flux via the following sequence of communications:

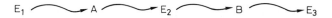

This is a sequence through which information is transmitted and hence is termed a regulatory sequence. In this case, it is the same sequence that transmits the flux (i.e. the transmission sequence for information is identical to that for flux). This is not necessarily the case. For example, the effector X communicates with the flux through the regulatory sequence:

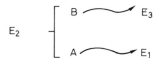

In this case, because of feedback regulation, some of the information flow is in the opposite direction to the flux

(i.e. $E_2 \longrightarrow A \longrightarrow E_1$).

It should also be noted that the intermediate, A, plays two quite distinct roles in these regulatory sequences; it is the pathway substrate for E_2 and is an allosteric inhibitor for E_1. The concentration of A will, therefore, depend on three factors: the flux J, the kinetic response of E_2 to A, and the kinetic (allosteric) response of E_1 to A. Consequently, in order to be able to define precisely the concentration of such a metabolic intermediate, it is necessary to know all the regulatory sequences in which it participates [12]. This applies, therefore, to an understanding of the factors that could be involved in the regulation of the concentration of an intermediate. That this is not a trivial point will be shown when the control of the concentration of second messengers is discussed below.

Regulatory Sequences and Second Messengers of Hormone Action. The concept of the second messenger was proposed by Sutherland et al. [20] in 1965 to account for the role of cyclic-AMP in mediating the intracellular response to a hormone. In this elegant scheme, the hormone is seen as the primary messenger which, after release from the endocrine gland and transport in the bloodstream, modifies metabolism within the target tissue(s) by changing the concentration of the second messenger. Since 1965, a considerable amount of work has been directed towards understanding both the molecular events by which a hormone interacts with a tissue and causes a change in the intracellular concentration of cyclic-AMP, and how the latter produces the physiological response in the target tissue. It is known that some hormones bind to a specific receptor on the surface of the cell, that this binding enhances the activity of adenylate cyclase, and that this leads to an increase in the steady state concentration of cyclic-AMP in the tissue. However, until recently, the precise quantitative roles of adenylate cyclase and the phosphodiesterases (which are the enzymes that produce and utilise cyclic-AMP, respectively) in the maintenance of the steady state concentration of this compound had not be considered. A number of models have now been proposed to account for both the maintenance and changes in the steady state concentration [1].

A simple model for the control of the concentration of a hypothetical secondary messenger is described below. The secondary messenger, X, is produced from compound A, via enzyme E_1, and X is utilised (i.e. inactivated) by conversion to B via enzyme E_2.

$$A \xmapsto{\;\;\;E_1\;\;\;} X \xrightarrow{\;\;\;E_2\;\;\;} B \xrightarrow{\;\;\;E_3\;\;\;} (J)$$

It should be obvious that this system is formally analogous to a metabolic flux and can be considered as such. The first important practical step in the investigation of the structure is to identify whether reactions are non- or near-equilibrium [14]. It is assumed that for most secondary messengers the reactions are in fact non-equilibrium. The second important step is to establish the flux-generating step and the regulatory sequence(s). The biochemical and practical significance of this can be appreciated by reference to the cyclic-AMP system. The question of whether adenyl cyclase catalyses a flux-generating step was investigated [1]: in all tissues investigated it appears to catalyse such a reaction (i.e. K_m for ATP $\ll$ intracellular concentration of ATP). This is of considerable importance since, if it did not catalyse a flux-generating step, or a regulatory reaction for the flux, as defined earlier, most of the in vitro properties of the cyclase would be physiologically meaningless. Thus, the concept of the regulatory sequence indicates that changes in the concentration of cyclic-AMP can only be produced via changes in the flux (i.e. effects on the flux-generating step) or changes in the kinetic properties of the utilising enzyme (i.e. the phosphodiesterases). Hence, in order to change the concentration of a second messenger, a hormone must affect either the flux-generating step for the 'second messenger pathway', or the properties (e.g. K_m or V_{max}) of the utilising system(s).

The regulatory sequence in the hypothetical model given above is $E_1 \rightsquigarrow X \rightsquigarrow E_2$. Although this simple model for the control of a secondary messenger provides a useful introduction for consideration of the principles of hormone action, it is an oversimplification. For many secondary messengers, including cyclic AMP, the utilisation process comprises at least two reactions, which have different properties. Nonetheless, a mathematical model can be constructed upon which the concentration of the secondary messenger can be calculated.

The following reaction sequence describes the model for cyclic-AMP:

The activities of adenylate cyclase and high and low K_m phosphodiesterase are represented by 1, 2 and 3, respectively. The following assumptions are implicit in the model: adenylate cyclase is saturated with substrate (i.e. it is a flux-generating step); all three enzymes catalyse non-equilibrium reactions; both phosphodiesterase enzymes obey Michaelis-Menten kinetics. Consequently, the model is described mathematically as follows:

$$V_1 = \frac{V_2 S}{K_{m_2} + S} + \frac{V_3 S}{K_{m_3} + S},$$

where S is the concentration of cyclic-AMP and subscripts 1, 2 and 3 refer to the enzymes as indicated above. Consequently, the concentration, S, can be calculated from the equation:

$$s = \frac{\sqrt{b^2 - 4ac} \pm -b}{2a},$$

where $a = (V_2 + V_3 - V_1)$, $b = (V_2K_{m_3} + V_3K_{m_2} - V_1K_{m_3} - V_1K_{m_2})$, and $c = (V_1 \cdot K_{m_2} \cdot K_{m_3})$.

Thus, the basal concentration of cyclic-AMP in a tissue can be calculated from the above equation from values of V_{max} measured in vitro for the three enzymes and the in vitro measured K_m values for the two phosphodiesterase enzymes. In addition, the increase in the concentration of cyclic-AMP produced by the action of a hormone can also be obtained if the effects of the hormone on the cyclase and phosphodiesterase enzymes are known. The relevant information would be included in the mathematical model given above. Such calculated concentrations can then be compared with the experimented concentration as a test of the validity of the model [1, 11]. Such an approach could now be attempted for other important second messenger-like regulators including cyclic-GMP, calcium ions, prostaglandins, adenosine and inositol trisphosphate.

Application of Control Principles to Branched Pathways

In practice, metabolic pathways are not always as straightforward as described above. They are usually branched, so that one flux divides into two or more fluxes which may also be further subdivided. In the following system:

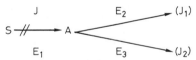

the flux generated at E_2 divides at A into J_1 and J_2 such that $J = J_1 + J_2$. This complexity influences the factors that affect the concentration of A and the relationship of A to the various fluxes. The concentration of A is now determined by the total flux, J, plus the kinetics of both enzymes E_2 and E_3; hence, the concentration of A is only partially determined by either flux J_1 or J_2. Consequently, although A is a simple internal effector for flux J it is only a partial internal (or partial external) effector for fluxes J_1 and J_2. (This shows that the terms 'internal' and 'external', when applied to effectors, must relate to a given flux and cannot be used in an absolute sense.) In this particular branched pathway, enzyme E_1 is regulatory for all three fluxes, J, J_1 and J_2, whereas, for example, enzyme E_2 is regulatory for J_1 and J_2 but not for J. Consequently, an effector of E_2 could change the fluxes J_1 and J_2 (in opposite directions) but J would not be affected. This re-emphasises the important point that to understand the control of flux, the *precise* flux under consideration must be defined [11, 12].

Sensitivity in Metabolic Regulation

Sensitivity in metabolic regulation can be defined as the quantitative relationship between the relative change in enzyme activity and the relative change in concentration of the regulator. (If the concentration of a regulator (x) changes by Δx, the relative change is Δx/x; similarly, if the flux (J) changes by ΔJ, the relative change is ΔJ/J. The sensitivity of J to the change in concentration of (x) is given by the ratio (ΔJ/J):(Δx/x) and this sensitivity is indicated by the symbol *s* [11, 12].) For example, if the concentration of a regulator increases twofold, the question arises, how large an increase in enzyme activity will this produce? The greater the response of enzyme activity to a given increase in regulator concentration, the greater is the sensitivity.

There are several important mechanisms for increasing sensitivity which include multiplicity of regulators, cooperativity (i.e. sigmoid response of enzyme activity to regulatory concentration) inter-conversion cycles and substrate cycles. These have been discussed in detail in several reviews [11, 14, 16]. However, attention will be focussed in this discussion on substrate cycles since we consider such cycles are particulary important in exercise [14].

Substrate Cycles. It is possible for a reaction that is non-equilibrium in the forward direction of a pathway (i.e. A → B) to be opposed by a reaction that is non-equilibrium in the reverse direction of the pathway (i.e. B → A). Both reactions must be chemically distinct (i.e. different reactions) so that they will be catalysed by separate enzymes. Then, a substrate cycle between A and B occurs if the two enzymes are simultaneously catalytically active. For every molecule of A converted to B and back again to A, chemical energy must be converted to heat, which is lost to the environment. An example of a substrate cycle is the fructose 6-phosphate/fructose bisphosphate cycle (fig. 2).

The role of a cycle can best be understood when it is appreciated that, in some conditions, an enzyme activity may have to be reduced to values closely approaching *zero*. Even with a sigmoid response this would require that the concentration of an activator be reduced to almost zero or that of an inhibitor to an almost infinite level. Such enormous changes in concentration probably never occur in living organisms, since they would cause osmotic and ionic problems and unwanted side reactions. However, the net flux through a reaction can be reduced to very low values (approaching zero) via a substrate cycle. Thus, as the concentration of the product of the

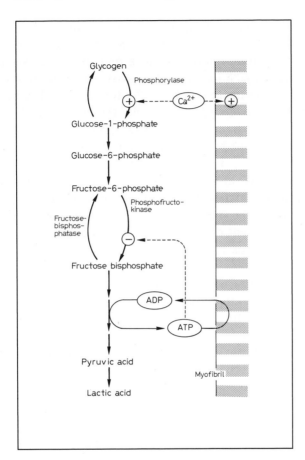

Fig. 2. There are at least two substrate cycles in muscle glycolysis: one between glycogen and glucose-6-phosphate and one between fructose 6-phosphate and fructose bisphosphate. The rates of both these cycles can be increased by catecholamines which should improve the sensitivity of these reactions to the external control agent, Ca^{2+} and ATP.

forward enzyme (i.e. fructose bisphosphate in the above example) reaches a low level it is converted back to substrate by the reverse enzyme (fructose bisphosphatase). This ensures that the net forward flux is very low despite a finite activity of the forward enzyme and a moderate concentration of an activator of 6-phosphofructokinase. Now if the concentration of this activator is increased by only a small amount above that at which the activities of the two enzymes are almost identical (and the flux is almost zero), the activity of 6-phosphofructokinase will increase so that the net flux through the reaction will increase from almost zero to a moderate rate. Such a cycle therefore provides a large improvement in sensitivity; indeed, it can be seen as a means of producing a threshold (or almost threshold) response with a simple metabolic system. Indeed, the cycle provides an improvement in sensitivity without changing the properties or characteristics of the enzyme catalysing the forward reaction in the pathway and, for this reason, it can be seen to be different from the other mechanisms for improving sensitivity; this has also been shown to be the case when the precise quantitative role of substrate cycles in metabolic control is considered [11]. One advantage of the substrate-cycling mechanism for increasing sensitivity is that the extent of this increase varies according to the rate of cycling, in other words sensitivity is proportional to the ratio, cycling rate/flux [14].

Considerable evidence has now been obtained to demonstrate this variability in cycling rates in different cycles.

(1) The precise rate of the fructose 6-phosphate/fructose bisphosphate cycle can be measured from the changes in the $^3H/^{14}C$ radioactivity ratio in hexose-monophosphates and fructose bisphosphate after isolation and separation from a tissue that has been utilising [5-^3H, 6-^{14}C]glucose. The hormone adrenaline, or other β-adrenoreceptor agonists, increases the cycling rate up to tenfold in the isolated epitrochlearis muscle of the rat. This stimulation occurred at physiological concentrations of the hormone and was abolished by the β-adrenoreceptor antagonist propanolol [6, 7]. The rate of the cycle was also increased after exercise [8].

(2) The precise rate of triacylglycerol/fatty acid cycle can be measured by comparing the rates of fatty acid and glycerol production by adipose tissue, or by following the rates of incorporation of [^3H] from ^3H_2O into the glycerol and fatty acid moieties of triacylglycerol. In isolated adipose tissue of the rat it has been shown that the cycling rate was increased markedly by β-adrenergic agents and the effect was abolished by the β-blocker propranolol [4]. Furthermore, the rate of this cycle in white adipose tissue of the mouse in vivo was doubled by feeding and increased fivefold by a β-adrenoreceptor agonist and, in brown adipose tissue, it was increased threefold by such an agonist and doubled by expose to the cold for 4 h [5]. Recent evidence has been obtained that this cycle occurs in adipose tissue in man – and that the rate of cycling can be increased tenfold in patients suffering from severe burns [22].

We can now illustrate how the substrate cycle can play a role in the regulation of flux by reference to the following example.

Controlling the Rate of ATP Generation in the Sprint

Every individual cell in the body, whether a nerve cell in the brain, a type IIB fibre in muscle or a spermatozoan, maintains a virtually constant ATP concentration. ATP is the energy currency of the cell and unless its concentration is maintained within narrow limits, the cell will suffer a 'cash-flow' problem. If services, such as maintaining the correct ion balance and removing water from the cell, cannot be paid for in terms of ATP hydrolysis they will not take place and the energy-bankrupt cell will die. Although the ATP concentration in the cell is small, it can be rapidly regenerated by increasing the rate of fuel oxidation, which means that as soon as the rate of ATP hydrolysis increases the rate of fuel oxidation must also increase; and, furthermore, the increase in the latter must match *precisely* the increase in rate of ATP hydrolysis. Let us see what this precision means for ATP production during sprinting in man.

At rest, the muscle fibre requires about 0.03 μmol ATP/s/g fibre but in sprinting the demand rises to about 3 μmol/s/g, a 100-fold increase. Because glycolysis now becomes the *sole* means of ATP generation (some of the ATP in the resting state will be provided from aerobic metabolism), its rate probably increases at least 1,000-fold, from that at rest, to provide the ATP required during sprinting. This poses a major problem in control. If the rate of glycolysis was understimulated by only 10% during sprinting, at the end of 10 s the ATP concentration in the fibres would have been reduced by half with serious consequences for the fibre. Of course, this cannot happen since, as the ATP concentration decreases, the cross-bridge cycle is progressively inhibited causing fatigue which effectively reduces the rate of ATP utilisation.

Only two reactions in the whole sequence of enzyme-catalysed reactions in glycolysis play a major role in the control of glycolytic flux (these are the regulatory reactions). The first step is the breakdown of glycogen, catalysed by the enzyme phosphorylase, and the second is third of the way along glycolysis, at a reaction catalysed by the enzyme phosphofructokinase. At these sites, two main regulatory agents are at work although a host of others play minor roles.

With typical parsimony, nature uses the same regulator not only to increase the rate of the cross-bridge cycling (i.e. contraction of the muscle) but also to increase the activity of phosphorylase and so increase the rate at which glycogen breaks down. This is an increase in the concentration of calcium ions. The concentration of ATP is the second important regulator of glycolysis. When the concentration of ATP falls, even by a very small amount, the enzyme phosphofructokinase works faster and glycolysis increases. In fact, the ATP concentration decreases very little but this causes changes in several factors that also affect the activity of phosphofructokinase (e.g. AMP, Pi, phosphocreatine, NH_4^+ fructose bisphosphate) (fig. 3). Conversely, if the concentration of ATP rises, the enzyme slows down. Such a negative feedback mechanism is ideal for maintaining a constant concentration, provided that the mechanism is sensitive to the controlling agent.

The supreme sensitivity of glycolysis to control by calcium ions and ATP and the other factors given above is rather surprisingly achieved by enzymes which catalyse reactions that oppose the specific glycolytic reactions. As a result, both forward and reverse reactions occur simultaneously. For example, phosphofructokinase catalyses the conversion of fructose 6-phosphate to fructose bisphosphate while the enzyme, fructose bisphosphatase catalyses the hydrolysis of fructose bisphosphate to fructose 6-phosphate. The simultaneous activity of the two enzymes will result in the cycling of fructose 6-phosphate to fructose bisphosphate and back (fig. 2). No net glycolytic reaction takes place yet ATP is hydrolysed to ADP and phosphate. This means that the chemical energy in ATP is converted into heat which is eventually lost from the body. For this reason the mechanism is sometimes termed a 'futile' cycle. But it is far from purposeless.

The role of the cycle is best explained by reference to the sprinter's behaviour immediately before an important 100-metre race. When the sprinter is resting in the dressing room before the race, the rate of glycolysis in the muscles will be low as will be the activities of both phosphofructokinase and fructose bisphosphatase. Consequently, the cycling rate and therefore the sensitivity of the system to changes in concentration of ATP will also be low. As the time for the race approaches, and particularly when the sprinter is on the blocks waiting for the gun, the increasing anxiety, nervous tension and anticipation will result in high levels of the stress hormones, adrenaline and noradrenaline. In the muscle, these hormones lead to a rapid activation of *both* phosphofructokinase and fructose bisphosphate. Although the flux through glycolysis remains low, since the race has not yet started, the rate of cycling is now high so that the system is very sensitive to small changes in concentration of ATP and the other regulators. These changes will occur as soon as the gun is fired.

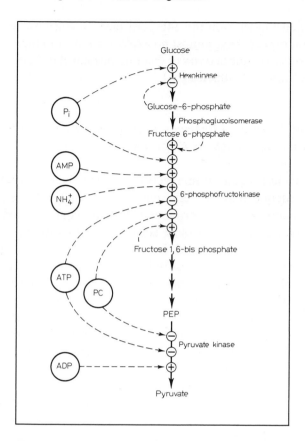

Fig. 3. There are three important sites for the regulation glucose conversion to pyruvate: hexokinase controlled by inhibition from glucose-6-phosphate and relief of this inhibition by phosphate; phosphofructokinase controlled by ATP inhibition which is potentiated by phosphocreatine (PC) and relieved by AMP, NH_4^+, fructose 1,6-bisphosphate, phosphate and fructose 6-phosphate; and pyruvate kinase which is inhibited by ATP and phosphocreatine and activated by ADP.

These sensitive control mechanisms ensure a smooth transition from the use of phosphocreatine to the use of glycogen to power the sprint. The importance of nervous tension in 100-metre sprinters is well known and was aptly emphasised by the trainer Sam Mussabini in the film *'Chariots of Fire';* Mussabini explaining to Harold Abrahams that he can beat Eric Liddell over the short sprint says 'A short sprint is run on nerves; it's tailor-made for neurotics.' In the 1924 Olympic Games, Abrahams won the 100-yard sprint and Liddell won the 440-yard race! In contrast, well-trained elite marathon runners have almost no explosive power in their leg muscles and, not surprisingly, the activity of fructose bisphosphatase is not detectable in muscle from such athletes [16].

Use of Maximum Activities of Enzymes as Quantitative Indices of Maximum Flux through Metabolic Pathways

The advantage of a near-equilibrium reaction, in a metabolic pathway in vivo, is that the reaction may be very sensitive to small changes in concentrations of co-substrate or co-product [11]. Consequently, large changes in flux can be transmitted through such a reaction without any requirement for complex regulatory properties. In general, this means that the activity of the enzyme can be measured relatively easily in crude extracts of the tissue; this ease-of-assay has, unfortunately, been used by some investigators as the only criterion for the selection of an enzyme to study the maximum flux through a metabolic pathway. This cannot be done. For example, the activity of glyceraldehyde 3-phosphate dehydrogenase has been used to indicate the maximum glycolytic capacity of muscle but the maximum activity of this enzyme can be an order of magnitude greater than the maximum flux through the pathway; the measured maximum glycolytic flux in muscles of the locusts, cockroach, honeybee and rat heart are 4, 15, 31, and 3.7, whereas reported activities of glyceraldehyde 3-phosphate dehydrogenase in these muscles are 330, 100, 150 and 240 µmol/min/g fresh muscle, respectively. A similar problem applies to the use of lactate dehydrogenase as a quantitative index of glycolysis, β-hydroxybutyryl-CoA dehydrogenase as a quantitative index of fatty acid oxidation, and fumarase, malate dehydrogenase or citrate synthase as quantitative indices of the citric acid cycle [9]. Metabolic logic tells us that these enzymes cannot be used as quantitative indices of flux but despite this they are *still* being used even in 1988. The theoretical discussion given above allows us to select the *key* enzymes than *can* be used for this purpose.

Enzymes that catalyse non-equilibrium reactions in a metabolic pathway provide directionality in that pathway and are usually subject to allosteric control (see above). Indeed, the control mechanisms may be complex. This means that knowledge of such control mechanisms must be available *before* a satisfactory assay method for measurement of the maximum activity can be developed; hence knowledge of metabolic control is necessary to enable the enzyme activity to be adequately assayed in crude extracts of the tissue (see below).

Theoretical Conditions for Quantitative Use of Maximum Enzyme Activities

There are at least two conditions that must be satisfied before an enzyme activity can be used to provide quantitative information. First, it is necessary to establish which enzymes in the pathway catalyse non-equilibrium reactions (see above). Secondly, it is necessary to demonstrate experimentally that the maximum activities of such enzymes in vitro can be used to indicate quantitatively the maximum flux through a reaction. This is done by comparison of the in vitro enzyme activity with the measured or calculated maximum flux through the pathway (table I). It is preferable if more than one species is used in such an investigation. The maximum flux through a pathway that involves oxidation of a fuel can be measured directly in the intact animal from arteriovenous concentration differences across an exercising muscle or indirectly from knowledge of oxygen uptake of the exercising muscle [16]. Alternatively, an indirect biochemical method can be used; the measurement of the maximum in vitro activity of myofibrillar ATPase in an extract of the given muscle; this method assumes that most of the ATP produced by the muscle will be consumed by this reaction during exercise. The rate of ATP utilisation can be used to calculate the rate in such pathways [23]. On the basis of the data presented in table I, it is considered that hexokinase provides an indication of the maximum flux from the pathway glycolysis-from-glucose, 6-phosphofructokinase or phosphorylase for the pathway, glycolysis-from-glycogen, and oxoglutarate dehydrogenase for the Krebs cycle (i.e. aerobic metabolism).

The physiological information that can be obtained from such studies will be illustrated with two examples, after a brief description of the practical aspects of enzyme activity measurement in crude extracts of tissues.

Practical Considerations

Tissue Disruption. The use of a ground-glass homogeniser fitted with a tight-fitting ground-glass pestle is the simplest method for tissue disruption. A small amount of tissue is weighed accurately, cut into small pieces

Table I. Maximum activities of hexokinase, phosphofructokinase, phosphorylase and oxoglutarate dehydrogenase and the measured or calculated maximum rate of carbohydrate utilization in selected muscle from some invertebrates and vertebrates

Animal	Muscle	Calculated rate of metabolic process in vivo to provide energy for muscle (µmol glucose or glucose equivalent min^{-1} g^{-1} at 25°C)			Enzyme activities (µmol min^{-1} g^{-1}) at 25°C, except where indicated			
		method	anaerobic glycolysis	complete oxidation of glucose	hexo-kinase	phospho-fructo-kinase	phosphor-ylase	oxoglutarate dehydro-genase
Great scallop (*Pecten maximus*)	catch adductor	myofibrillar ATPase	1.9	0.15	0.2	4.5	1.8	–
Penwinkle (*Littorina littorea*)	pedal retractor	myofibrillar ATPase	3.5	0.28	0.8	12.4	7.8	–
Thick top shell (*Monodonta lineata*)	pedal retractor	myofibrillar ATPase	10.0	0.79	1.0	10.9	3.8	–
Common limpet (*Patella vulgata*)	radular retractor	myofibrillar ATPase	3.9	0.31	3.1	13.3	3.0	0.9
Common squid (*Loligo forbesi*)	fin	myofibrillar ATPase	11.4	0.90	0.4	10.5	10.8	–
Lobster (*Homarus vulgaris*)	claw adductor	myofibrillar ATPase	9.2	0.73	0.6	6.0	4.4	–
Locust (*Schistocerca gregaria*)	flight	O$_2$ uptake of flying insect	178	14.0	11.5	7.5	17.0	12.0
Cockroach (*Periplaneta americana*)	flight	O$_2$ uptake of flying insect	190	15.0	18.0	19.0	30.0	27.0
Honey bee (*Apis mellifera*)	flight	O$_2$ uptake of flying insect	406	32.0	29.0	23.0	4.0	23.0
Blowfly (*Lucilia sericata*)	flight	O$_2$ uptake of flying insect	749	59.0	35.0	43.0[a]	55.0[a]	23.0
Silver-Y moth (*Plusia gamma*)	flight	O$_2$ uptake of flying insect	165	13.0	50.0	41.0	2.0	13.0
Dogfish (*Scylliorhinus canicula*)	flight	myofibrillar ATPase	3.8	0.3	0.5	4.2	3.5	–

Animal	Muscle	Basis of measurement						
Cod (*Gadus morhua*)[b]	red	myofibrillar ATPase	5.3	0.3	–	4.0	–	–
Herring (*Clupea harengus*)[b]	red	myofibrillar ATPase	2.3	0.5	0.1	1.4	2.1	–
Flounder (*Paraiichthys flessus*)[b]	red	myofibrillar ATPase	3.3	0.3	0.1	3.9	–	–
Bass (*Dicentrarcus labrox*)[b]	red	myofibrillar ATPase	2.1	0.2	0.2	5.6	9.8	–
	white		11.6	0.9	0.2	13.3	15.3	–
Mackerel (*Scomprus scomprus*)[b]	red	myofibrillar ATPase	1.9	0.2	0.7	4.8	4.6	–
	white		10.0	0.8	–	6.6	8.8	–
Trout (*Salmo gairdneri*)	red	O$_2$ uptake during continuous swimming	15.2	1.2	2.6	12.2	14.0	1.0
Frog (*Rana temporaria*)	sartorius	rate of lactate production by isolated muscle	13.0	–	1.3	22.0	29.0	–
Pigeon (*Columba livia*)	pectoral	O$_2$ uptake of flying bird	48	3.8	3.0	24.0	18.0	2.1
Rat (Wistar strain)	heart	perfused working in Xvitro	47	3.7	6.1	10.0	12.0	5.5
Human (normal)	quadriceps	lactate production during sprinting	35	–	–	28.5	–	–
Human (long-distance runner)	quadriceps	O$_2$ uptake during marathon running		0.5	0.5[c]	–	–	–

Experimental details of enzyme assays, the sources of published metabolic rates in intact animals or intact muscles and the assumptions made in converting these rates to μmoles of fuel utilized per minute per gram of muscle are described in references 10 and 19. The ranges of enzyme activities are also given in these references. When myofibrillar ATPase activity is used for calculating the rate of carbohydrate utilization it is assumed that anaerobic glycolysis (from glycogen) produces 3 ATP molecules per glucose residue utilized so ATPase activity is divided by 3.0, whereas for the complete oxidation of glucose 38 ATP molecules are produced so that ATPase activity is divided by 38. The activities of 2-oxoglutarate dehydrogenase have been divided by two so that they are comparable with the other activities and with the calculated rates (one molecule of glucose gives rise to two molecules of acetyl CoA for oxidation in the cycle).

[a] Enzyme activities measured on muscles from *Calliphora erythrocephala*.
[b] Enzyme activities (including myofibrillar ATPase) measured at 10 °C and results presented at 10 °C rather than 25 °C.
[c] Mean of activities in biopsy samples from three individual long-distance runners.

and placed in a homogeniser (cooled on ice). Homogenisation medium is added and the contents homogenised manually, by a firm, predominantly rotary action of the pestle; vertical movements of the pestle should be made carefully since this can cause frothing and hence inactivation of enzymes. When homogenisation is complete the homogenate is kept on ice until enzyme activity assay. The glass homogeniser can be replaced by the Polytron which is particularly useful for homogenisation of muscle from vertebrates and non-insect invertebrates. Even after a thorough homogenisation, mitochondria may not have been completely disrupted, or vesicles may have been formed from intracellular membranes with enzyme trapped within them. The homogenate then can be further treated by ultrasonic vibrations (sonication – usually for periods of up to 1 min) or, alternatively, detergents can be used: deoxycholate or Triton X-100 is mixed with the homogenate (e.g. 10 μl of a 10% (w/v) solution of deoxycholate is added to each ml of homogenate). Triton X-100 may also be mixed with the assay buffer in the incubation tube or cuvette. An advantage of adding the detergent to the homogenate is that the final detergent concentration in the assay can be maintained very low (0.1%). However, detergents can inactivate enzymes and the effect of various detergents and various concentrations should be studied on the activity of the enzymes before a detergent is finally included in the extraction or assay medium.

Protection of Enzyme Activity during and after Homogenisation. It is necessary to buffer the homogenising medium to achieve a pH at which the enzyme is most stable. The initial pH of the buffer may change upon homogenisation of the tissue due to acid formation; many enzymes are most stable at neutral pH but some are stable at other pHs (e.g. 6-phospho-fructokinase is most stable at pH 8.2). The buffer concentration is usually 20–100 mM.

Tissue homogenisation causes a dilution of the enzymes, which may inactivate them. This can be minimised by addition of protein (usually bovine serum albumin at a final concentration of approximately 0.1–1% (w/v) or a high concentration of glycerol) to the homogenisation medium. Heavy metal ions inhibit many enzymes; they can be introduced into a homogenate from the tissue, from glassware, from distilled water or as a contaminant in commercial reagents. Interference may be reduced or prevented by inclusion of a chelating agent (usually 1–2 mM EDTA) in the homogenisation medium. The oxidation of the thiol groups of enzymes can cause inactivation; inclusion of a thiol reagent in the homogenising medium (e.g. 2-mercaptoethanol, dithiothreitol, N-acetyl cysteine) overcomes the problem.

The following medium has been found to be suitable for the extraction of many enzymes from several tissues, including muscle, brain, liver and mammary gland: 50 mM triethanolamine-KPH, 2 mM MgCl$_2$, 1 mM EDTA, 2 mM dithiothreitol at pH 7.5. Provided that subsequent assays are not affected, the buffer may be neutralised with NaOH, and dithiothreitol may be replaced by 30 mM mercaptoethanol.

Whenever possible, tissue should be taken from a freshly killed animal (or removed from a living animal by biopsy) and should be used as soon as possible after removal. However, when this is not practical (e.g. if tissue is obtained from an abattoir), the tissue should be stored and transported in ice if the delay is relatively short, but it should be frozen, preferably at liquid nitrogen temperatures (approximately 196 °C) if the period between obtaining the sample and carrying out the assay is long (e.g. 6 h).

Method of Assay of Enzyme Activity. Any quantitative technique of chemical analysis can be the basis of an enzyme assay. Techniques that have been used range from mass spectrometry to counting bubbles of gas. Assays can either be continuous, in which case the reaction is measured as it is taking place, or discontinuous, in which samples of the reaction are taken at intervals and assayed after termination of the reaction.

The spectrophotometric assay is probably the most widely used of all assay methods; it depends upon a difference in the amount of light that is absorbed by substrate and products at a given wavelength. The reaction is carried out in a spectrophotometer and changes in absorbance which are proportional to concentration are recorded continuously. Reactions involving the production or uptake of hydrogen ions can be assayed by the use of a glass electrode in a pH stat. The enzyme is incubated with substrate in an unbuffered medium and, as the pH changes, acid or base is automatically added to maintain a constant pH. The volume added is recorded and used to calculate the extent of the reaction. Reactions in which gas is produced can be followed by measuring the changes in pressure occurring when the reaction takes place in a closed vessel. The method is semicontinuous in that readings are taken at intervals but without stopping the reaction. Enzymes in which molecular oxygen is a reactant can often be conveniently assayed with an oxygen electrode.

For sampling assays, special consideration must be given to the means of stopping the reaction when samples are taken. Two important criteria are that termination must be virtually instantaneous and that it must not interfere with subsequent analytical procedures. With a new assay it is advisable to test that the reaction has been stopped completely. Several means of terminating reactions are available. Heat is widely used and is frequently effective (it is usually necessary to boil the sample and care should be taken that heat transfer is rapid, since the catalytic reactions will increase in rate as the temperature rises until denaturation occurs), strong acids or bases denature enzymes (final concentrations of about 3% (w/v) are particularly effective); and water-miscible organic solvents (e.g. ethanol, acetone) have also been used (the proportion of the solvent required to stop the reaction must be determined, but it is likely to be at least 50%). This list of stopping reagents is not exhaustive and any agent including a specific inhibitor, known to inhibit an enzyme can, in principle, be used to stop the reaction.

In a radiochemical assay, a radioactively labelled substrate is provided and the rate of appearance of radioactive product determined. Such methods depend upon the complete separation of product from substrate. The separation of substrate and product is at the heart of a radiochemical assay and is also likely to be the most time-consuming step. Chromatography provides the basis for most separations and ion-exchange methods have proved particularly useful. Separations can be simplified if an 'all or nothing' procedure is possible. The mixture is applied to an absorbent (usually an ion-exchange material) which is then washed to remove either all the product or all the substrate. A simple and widely used application of this principle is to use ion-exchange paper discs. For example, hexokinase can be assayed by using [^{14}C]glucose as substrate and applying samples (after stopping the reaction) to discs of DEAE-cellulose paper. These are then washed with water which removes the [^{14}C]glucose but the product (glucose-6-phosphate) is retained on the positively charged paper. The latter is dried and the radioactivity is measured after placing the disc in scintillation fluid. Small columns of ion-exchange resin can be produced using Pasteur pipettes and they can be used to separate substrate and product. In this case, the substrate is retained by the column and the product is not retained. Hence, the eluate is collected and its radioactivity estimated. In assays based on an 'all or none' separation, it is essential to establish that no products of side reactions separate with the product of the enzyme under study.

An advantage of such radiochemical assays is that, provided the counting efficiency is constant, there is no need to determine the absolute radioactivity (dpm) of the samples, since only the ratios of radioactivity are used. Two further measurements are, however, required: the total radioactivity present in a sample (determined without separation) and the total amount of substrate (in µmol) present in the sample. From these last two measurements, the specific activity of the substrate can be calculated [17].

Comparison of the Maximum Rates of ATP Formation via Anaerobic and Aerobic Metabolism in Muscle

Since the conversion of one glucose-equivalent from glycogen to lactate produces three molecules of ATP, the maximum rate of ATP formation from anaerobic glycolysis can be calculated by multiplying the maximum activity of 6-phosphofructokinase (measured as described above) by three. Similarly, the maximum rate of ATP formation from oxidation of glucose via the Krebs cycle and electron transport can be calculated by multiplying the maximum activity of oxoglutarate dehydrogenase by 18 (assuming that the complete oxidation of one molecule of glucose produces 36 molecules of ATP). (The factor is similar for the complete oxidation of a fatty acid molecule via the Krebs cycle.) Consequently, from the activities of 6-phosphofructokinase and oxoglutarate dehydrogenase, the *maximum* rates of ATP generation from anaerobic and aerobic metabolism, respectively, can be calculated and the maximum capacities of these two important ATP-producing systems compared (table II). The maximum rate of ATP formation from anaerobic glycolysis is about 2-fold greater in the pectoral muscle of the domestic fowl than that of the pigeon. But the maximum rate of ATP formation from the Krebs cycle, as indicated by oxoglutarate dehydrogenase, is almost 27-fold greater in the pigeon pectoral muscle compared to that of the domestic fowl (table II). In domestic fowl pectoral muscle, the maximum rate of ATP formation via the Krebs cycle could provide only about 1 % of that which could be provided from anaerobic glycolysis. This is consistent with the use of this muscle exclusively for short-lived maximum exercise during escape, which is characteristically powered by anaerobic metabolism. In contrast, the Krebs cycle is important during sustained flight in the pigeon, which can fly several hundred miles (non-stop) and hence requires the 'efficient' ATP formation of aerobic metabolism. A similar comparison is also done for muscles of trained and untrained man and woman (table III).

Table II. Maximum activities of 6-phosphofructokinase and oxoglutarate dehydrogenase and calculated maximum rates of ATP formation in muscles from domestic fowl, pigeon, rat and man [for source of data and bases of calculation, see ref. 3]

Animal	Muscle	6-Phospho-fructokinase	Oxoglutarate dehydrogenase	Calculated maximal rate of ATP formation from glucose or glycogen (μmol min^{-1} g^{-1} fresh weight at 25 °C)	
				anaerobic glycolysis	oxidation via Krebs cycle
Domestic fowl	heart	14	4.0	42	72
	pectoral	76	0.8	228	3
Pigeon	heart	16	3.1	48	56
	pectoral	46	4.8	138	86
Rat	heart	9.9	6.4	30	115
	diaphragm	22	3.2	66	58
	soleus	10	1.0	30	18
	gastrocnemius	42	1.1	126	20
Man (male well-trained)	vastus lateralis	24	1.4	72	25

Table III. Calculated maximum rates of ATP formation in human muscle based on maximal activities of key indicator enzymes under aerobic and anaerobic conditions

Group	Sex	Calculated maximum rate of ATP formation from glucose or glycogen (μmol min^{-1} g^{-1} fresh weight at 25 °C)	
		anaerobic glycolysis	oxidation via Krebs cycle
Untrained	male	104	13
	female	87	16
Medium-trained	male	91	21
	female	89	19
Well-trained	male	72	26
	female	61	29

Maximum rates of ATP formation are calculated as follows: for anaerobic glycolysis 6-phosphofructokinase activity is multiplied by 3; for oxidation by the Krebs cycle, oxoglutarate dehydrogenase activity is multiplied by 18 [data from ref. 3].

Metabolic Fuel Interrelationships

The body always has a demand for energy which is normally met from the food we eat. However, eating is not a continuous process, so that the body has to store energy for use between meals, during more prolonged starvation or when a large amount of extra energy is needed as in endurance running. The two main energy stores in the human body are glycogen, a carbohydrate, which is stored in muscle and liver, and fat, which is stored in adipose tissue. Glycogen as a fuel for exercise has been discussed in this volume.

The second major storage fuel in humans is fat, which is composed of triglyceride molecules. Triglyceride is of quite a different chemical nature to glycogen, being a much smaller molecule and containing proportionally less oxygen. The major store of triglyceride is in adipose tissue, which is composed of cells called adipocytes, each of which contains a large droplet of triglyceride, which occupies some 90% of the cell. Unlike most other tissues, adipose tissue does not form a discrete organ but is widely distributed in separate depots throughout the body. One mechanical principle adhered to by the body is that the largest amount of adipose tissue occurs near to the centre of gravity where it interferes less with locomotion.

Making the Fuels Available

Neither glycogen nor triglyceride can traverse the membrane of the cell in which they are stored and so must be broken down to smaller units before transport, glycogen to glucose and triglyceride to glycerol plus fatty acids.

The conversion of glycogen to glucose involves hydrolysis, that is, splitting by addition of water: this does not occur directly but first involves attack by a phosphate ion which splits off a terminal glucose unit as glucose-1-phosphate. The reaction if catalysed by the enzyme phosphorylase. In order to provide glucose as a fuel for muscle during very sustained exercise (as in the marathon or ultramarathon run), the phosphate must be removed and the glucose then traverses the liver cell membrane and enters the bloodstream and, hence, maintains the blood glucose concentration. This process is controlled by hormones including adrenalin, glucagon and vasopressin.

The hydrolysis of triglyceride to glycerol and fatty acids is a simpler process, catalysed by the enzyme triglyceride lipase which is controlled by hormones including insulin and the catecholamines. The fatty acids are transported across the cell membrane of the adipocyte into the blood in which they are carried to the muscle or other tissues. The transport of fatty acids poses a problem since they are not at all soluble in the blood plasma. The problem is overcome by combining the fatty acid molecule with albumin, a soluble protein present in the plasma. The binding is fairly tight but, as the blood passes through a working muscle, the fatty acid-albumin complex dissociates into albumin and fatty acids and the latter diffuse into the muscle fibre. This occurs because the concentration of fatty acids within the muscle is very low due to their removal by the process of oxidation.

How Much Fuel to Store and Why?

The amounts of stored glycogen and triglyceride are far from equal, as can be seen from table IV. In fact, some 98% of energy reserves are held in the form of triglyceride, enough to tide the average man over several weeks of starvation. Why should this be? The simple answer is that fat is a far more efficient fuel for storage. In a storage fuel, the ratio of its energy content to its mass is of paramount importance. The energy content of each of the two fuels can be compared by burning them in a calorimeter in which all the available energy is released as heat that is then measured. In the calorimeter, the fuels undergo the same overall chemical reaction (and therefore produce the same amount of energy) as they would if oxidised in a cell. Burning 1 g of pure glycogen releases 16 kJ, whereas burning 1 g of a typical triglyceride releases 35 kJ. Clearly, this establishes triglyceride as the more efficient storage fuel but the difference becomes even more pronounced when it is realised that, in the cell, triglyceride is stored in a pure state but glycogen is not. Glycogen, with its branched structure (necessary for rapid breakdown) and multitude of hydroxyl groups, is inevitably associated with a large number of trapped water molecules. More than half of the mass of the stored glycogen is composed of water and when this is taken into account, triglyceride emerges as the better storage fuel by a factor of at least five. The importance of storing triglyceride is emphasised by a simple calculation; if a 70-kg man stored glycogen instead of fat, for the same energy reserve, his weight would become about 105 kg!

Why, then, is glycogen stored at all? There are at least three reasons. First, the brain requires a constant supply of glucose, since it cannot oxidise fatty acids, and liver glycogen provides a readily available reserve for this vital tissue. Secondly, tissues may need to generate ATP under conditions when the amount of oxygen present is inadequate for complete fuel

Table IV. Fuel stores in the average man

Fuel stores	Fuel reserves		How long do the reserves last?, min	
	g	kJ	walking[1]	marathon running[1]
Adipose tissue triglyceride	9,000	337,500	15,500	4,018
Liver glycogen	100	1,660	86	20
Muscle glycogen	350	5,800	288	71
Blood glucose	3	48	2	< 1

[1] It is assumed that the energy expenditure during walking is about 22 kJ per minute and during marathon running (elite runner) is 84 kJ per minute. The amount of adipose tissue triglyceride will be much less in many elite marathon runners (perhaps 4,000 g) [see ref. 16 for details].

oxidation; glucose (that is, glycogen), but not fat, can produce ATP in the absence of oxygen. Thirdly, the rate at which glycogen can be oxidised by muscle is *greater* than that for fat (see below) – hence the rate of energy formation is greater and therefore the power output, that is the pace, is greater. This is of immense importance to the athlete!

If fat was the sole fuel used for the marathon, a simple calculation shows that about 300 g would be used during a race, so that the total stores would provide energy to satisfy the demands of even the elite runner for more than 3 days and 3 nights of running. So is fat used during the run, and if so, why does exhaustion occur so soon? In brief, the answer is that fat is used but, if it was the sole fuel, it could not provide energy at the high rate demanded by the marathon runner. The evidence for the latter is as follows [16]. First, if the carbohydrate stores of body are depleted prior to the run, for example, by eating a low carbohydrate diet for several days, exhaustion occurs earlier in a run. This is an experiment a runner can readily perform; obtain from the physician or dietician a diet that provides less than 25 g of carbohydrate each day, but enough fat and protein to produce satiety. Follow the diet for 3 days and note running performance. Secondly, in very long runs (e.g. > 100 miles) the power output declines to a level where the oxygen consumption is about half of that achieved maximally; at this stage very little carbohydrate is left in the muscle or liver so most of the energy *must* be derived from fat oxidation, implying that this could provide energy at only half of the maximal aerobic capacity. Thirdly, studies on a patient with a deficiency of the enzyme phosphofructokinase in muscle, who was, therefore, unable to carry out glycolysis and hence glucose oxidation and so was totally dependent upon fatty acid oxidation, showed that his maximum oxygen consumption was only 60% of that expected [Cerretelli, personal commun.]. The general conclusion is, therefore, that fatty acids are used during aerobic exercise but that the rate at which they can be used is limited to *about* 50% of the maximum rate of energy generation that is required to power the muscles; the difference must be made up by oxidation of carbohydrate (mainly muscle glycogen).

Switching Fuel Utilisation

If the above reasoning is correct, the marathon runner would achieve his best performance employing the following strategy. Use fatty acids from the very beginning of the rate at the maximum rate consistent with safe transport and make up the deficit in energy production by oxidation of carbohydrate (mainly muscle glycogen) at a rate determined by requirements. However, although fatty acids will be mobilised from adipose tissue, their concentration will not exceed 2 mM and this is lower than that of glucose [16], so, will not muscle prefer to use glucose in preference to fatty acid, especially as the runner will probably have been on a high-carbohydrate diet before the race? The answer is no because of a very sophisticated control system which ensures that when glucose and fatty acids are both available to the muscle, the latter are used in preference but any deficit in energy production is made up by glucose oxidation. This control mechanism is described below.

If the concentration of ATP in the muscle fibre falls below a 'pre-set' level, as it will when muscle contraction commences, this sets in train a series of events which increase the catalytic activity of key enzymes in glycolysis. These are the enzymes that have previously limited the flow rate. Conversely, a rise in ATP concentration will reduce the flow-rate. Such negative feedback serves to maintain the concentration of ATP within narrow limites (± 25%) despite very large changes in its rate of utilisation. In present context, the important point is that fatty acid oxidation increases the effectiveness of ATP as a feedback inhibitor of glycolysis and hence glucose and glycogen utilisation are decreased (fig. 4). Two important features of this control are worthy of note. First, since the rate of fatty acid oxidation is determined by the fatty acid concentration in the blood and this in turn is determined by the rate of fatty acid release from adipose tissue, it follows that this latter process can actually exert some

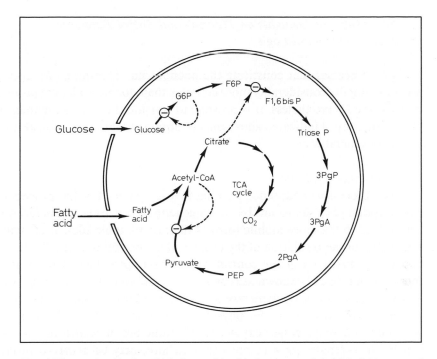

Fig. 4. The mechanism by which fatty acid oxidation in muscle inhibit the process of glycolysis. Increased rates of fatty acid oxidation raises the concentrations of acetyl CoA and citrate. Inhibition of pyruvate dehydrogenase and phosphofructokinase results from elevated concentrations of acetyl CoA and citrate, respectively. Inhibition of pyruvate dehydrogenase reduces the rate of oxidation by the TCA cycle. Inhibition of phosphofructokinase by citrate raises the glucose-6-phosphate level which inhibits hexokinase. Hence, the rates of glycolysis *and* glucose oxidation are decreased by fatty acid oxidation [16].

control over the rate of carbohydrate utilisation in the body. Secondly, although fatty acid oxidation reduces the rates of glycolysis and pyruvate oxidation, these processes still operate more rapidly than at rest. Indeed, the mechanism is such that if the demand for energy by the muscle increases, despite the oxidation of fatty acids, carbohydrate utilisation *will* increase to meet this demand. This increased rate will be maintained until the energy demand is decreased or until the body runs out of carbohydrate (see below). This control interrelationship between carbohydrate and fatty acid is known as the glucose/fatty acid cycle and has been discussed in detail elsewhere [16].

*Some General Thoughts on Training and Ageing Based on
Metabolic-Control-Logic*

The processes that constitute the phenomena of training and ageing
are undoubtedly considerably more complex than, for example, the process
of glycolysis. Nevertheless, it is possible to consider how the principles of
metabolic control indicated above may be applied to these systems, at least
in a very general way.

Endurance Training

The process of endurance training involves control of the process of
transcription [21]; the means by which genetic information in the DNA is
transferred to the more mobile form of the mRNA molecules. It is tacitly
assumed that the regulation of the process of transcription is totally different to that involved in the control of metabolic processes. This control
mechanism is usually known as repression, i.e. inhibition of the activity of
DNA-dependent RNA polymerase towards part of the genome. Repression
is considered to involve the physical restriction of these parts of the
genome to the polymerase enzyme. Non-histone proteins, histones or special RNA molecules (or a combination of both) may be involved in this
physical restriction. Such molecules are known as repressor molecules. The
presence of the repressor prevents the formation of mRNA for the synthesis of specific proteins, so that these proteins are lacking in cells when this
repressor is active: removal of the repressor permits that portion of the
genome to be transcribed so that new specific proteins will be produced.

Let us assume that the repressor is a protein molecule that binds to a
particular regulatory site (operator site) on the genome and prevents the
binding of the polymerase enzyme so that transcription of a portion of the
genome cannot take place. The process of development requires that this
part of the genome should be totally occluded, so that mRNA synthesis
relevant to this part of the genome cannot occur. How can this be
achieved? The repressor protein must bind to the genome so tightly that all
the available sites are fully occupied. One way of achieving such tight
binding is by a covalent attachment to the genome which would require a
specific enzyme. In this case, removal of the repressor action could *only* be
achieved by a specific cleavage of this covalent bond, i.e. a separate enzymatic reaction would be required. The enzyme would need to be totally (or
almost totally) inactive in the repressed state whereas it would need to be
highly active in the de-repressed state. Such large changes in activity could

only be achieved by an interconversion cycle, by a substrate cycle or possibly a substrate cycle combined with an interconversion cycle. On the other hand, the repressor molecule could bind reversibly to the DNA but, in order to achieve total occupancy of all the sites, the binding constant would have to be very high. However, a problem then arises during derepression. In order to facilitate dissociation of the repressor molecule from the DNA, the concentration of repressor molecule must fall to *very, very* low levels. It is possible that such large change in concentration of the repressor could only be achieved via a cyclic flux, analogous to a substrate cycle, i.e. the concentration of the repressor will be controlled both by a 'formation' reaction and a 'utilisation' reaction, as follows:

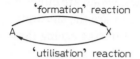

Since the net flux across this cycle is zero it is possibly more analogous to an interconversion cycle discussed above and it could result in very large but controlled changes in the concentration of the repressor, provided that there were adequate controls of the rates of both the formation and utilisation reactions. A similar cyclic system to this has been described for the production of large changes in the concentration of adenosine, an important regulator of blood flow in muscle and brain and possibly an important modulator of neurotransmission in nervous tissue [1].

Ageing and Performance

At the molecular level, the process of ageing is not understood. There is no doubt that ageing is characterised by structural changes in some (possibly many) proteins (e.g. collagen, elastin). The structural changes in the protein may be caused by errors in the translation process, but it is also possible that these structural changes may occur in the process of transcription so that they will be incorporated into the mRNA. The central dogma of molecular biology is that information is transferred in the direction,

$$DNA \rightarrow mRNA \rightarrow protein.$$

Consequently, any errors that occur in the synthesis of DNA or mRNA will find their way into the proteins and this could result in changes in the structure of cell constituents and changes in enzyme activity. It is likely that such changes will be detrimental to the well-being of the animal. So far, however, there are only one or two known examples of changes in protein structure which can explain some of the very obvious macroscopic changes that characterise ageing (e.g. changing structure of the protein of the lens). The following discussion contains a speculative and provocative hypothesis which attempts to provide a molecular explanation for some of the macroscopic changes that characterise ageing and consequently affect athletic performance.

Failure of Control Systems during Ageing. The improvement in sensitivity in the regulation of activity of enzymes that catalyse non-equilibrium reactions is achieved through a limited number of mechanisms (see above). Two of these mechanisms could be described as dynamic, in that they depend upon the kinetics of one or more enzymes, which are either mainly or solely involved in regulation. These mechanisms are the substrate cycle and the interconversion cycle. It is suggested that during ageing these two separate cycles progressively fail to provide the increased sensitivity that is required for normal metabolic control.

The role of these cycles is to provide an improvement in sensitivity of the regulatory process, so that when a concentration change in a regulator occurs, the change in enzyme activity will be sufficiently large to meet the demands of the physiological situation. The regulator(s) could be operative within the cell (e.g. the change in the adenine nucleotide levels during exercise) or it could be external to the cell (e.g. a change in the concentration of a hormone which mediates its effect through the intracellular secondary messenger (see above). The rate of the interconversion cycle or the rate of the substrate cycle will provide the necessary degree of amplification. However, if the maximum rates of these cycles are reduced during ageing, the sensitivity to the change in regulator could be very *markedly* reduced. There are at least three consequences of such a reduction in sensitivity.

(i) If the change in regulator concentration cannot provide a sufficient response from the metabolic system, then the physiological response will not be adequate or satisfactory. This can best be illustrated by an example: the energy for the contractile process in sprinting is provided primarily from glycolysis, and the non-equilibrium enzymes in this process respond to changes in the concentration of certain regulators. If the sensitivity of the control mechanism is reduced, then the increase in the rate of glycolysis will not be sufficient to meet the energy demands of the contractile process. The result of this will be that the rate of energy utilisation by the contractile process must be curtailed, so that the acceleration in the sprint will be reduced. This is, of course, a well-known phenomenon of ageing.

(ii) A second problem is that in an attempt to increase the rate of the metabolic process in relation to the physiological response, the increase in the concentration of regulator may be much larger than is normal. Such an increase in the regulator concentration would be necessary to overcome the lack of sensitivity of the regulatory process. It is, however, possible that a very large increase in the concentration of a regulator could be detrimental to the physiology of the particular tissue: large changes in the concentration of metabolites can disrupt the organised kinetic structure of a metabolic process and, in addition, a large change in the concentration of a reactive intermediate could lead to the accumulation of side products which could be detrimental and even dangerous to the well-being of the cell. (For example, a large increase in the accumulation of intracellular glucose in a muscle in an attempt to stimulate glycolysis and energy production could lead to the accumulation of the less readily metabolised sugars, fructose and sorbitol, via the polyol pathway. There is indeed evidence that the accumulation of fructose and sorbitol may be responsible for damage to nerves and possibly to the lens in diabetes mellitus [16]).

(iii) Although in the above discussions substrate cycles were discussed as a mechanism for regulation of the flux through metabolic pathways, they can also be involved in the regulation of the precise concentration of such things as intracellular ions, for example, sodium and calcium ions. A reduction in the rate of such cycles could decrease the precision of the regulation so that large fluctuations in the intracellular concentrations of such ions could result. This could cause detrimental changes in the rate of many important processes within the cell and, moreover, it could result in ionic and osmotic imbalance, which could eventually lead to cell death.

It is possible that the capacity for both interconversion cycles and substrate cycles is progressively reduced with age. In other words, the concentration of the enzymes which limit the rate of operation of these cycles would be decreased during ageing. Alternatively, or perhaps in addition, it would be possible for the rate of such cycles to be reduced by an impairment in the mechanism that responds to the change in the concentration of the regulator. This could be as simple as a decrease in the affinity of an allosteric site on the enzyme that controls the rate of the cycle, or it could be as complex as an impairment in the receptor for a hormone on the membrane of the tissue. (The effect of binding the hormone to the receptor is to cause a change, eventually, in the activity of the enzyme that controls the rate of the substrate or interconversion cycle). In either event the effect will be a reduction in the sensitivity of the regulatory system.

Since a large number of processes in all tissues of the body may be controlled by interconversion or substrate cycles, then a gradual impairment in the rate of such cycles with age could lead to a generalised lack of responsiveness of tissues and physiological processes within those tissues. This would result in a general decline in what might be termed the 'efficiency' of the many and varied physiological processes in the tissues and organs of the body.

The author has discussed elsewhere how a reduction in the rate of substrate cycles could, in particular, account for obesity and, since this is common in middle and old age, it would be consistent with the ideas proposed above [18]. Furthermore, the syndrome of accidental hypothermia, which is very prevalent in debilitated elderly people in no more hostile an environment than their own home, could also be explained by the inability to increase the rate of substrate cycles in response to hypothermia. The energy that is normally expended in such cycles is released as heat and this may be one process by which heat can be generated in warm-blooded animals. A marked decrease in the capacity of such cycles in elderly people would preclude the possibility of heat generation by a stimulation of such cycles.

References

1   Arch, J.R.S.; Newsholme, E.A.: The control of the metabolism and hormonal role of adenosine. Essays Biochem. *14:* 82–123 (1979).
2   Blomstrand, E.; Challiss, R.A.J.; Cooney, G.J.; Newsholme, E.A.: Maximal activities of hexokinase, 6-phosphofructokinase, oxoglutarate dehydrogenase and carnitine palmitoyltransferase in rat and avian muscles. Biosci. Rep. *3:* 1149–1153 (1983).
3   Blomstrand, E.; Ekblom, B.; Newsholme, E.A.: Maximum activities of key glycolytic and oxidative enzymes in human muscle from differently trained individuals. J. Physiol. *381:* 111–118 (1986).
4   Brooks, B.J.; Arch, J.R.S.; Newsholme, E.A.: Effects of hormones on the rate of the triacylglycerol/fatty acid substrate cycle in adipocytes and epdidymal fat pads. FEBS Lett. *146:* 327–330 (1982).
5   Brooks, B.J.; Arch, J.R.S.; Newsholme, E.A.: Effect of some hormones on the rate of the triacylglycerol/fatty substrate cycle in adipose tissue of the mouse in vivo. Biosci. Rep. *3:* 263–267 (1983).
6   Challiss, R.A.J.; Arch, J.R.S.; Newsholme, E.A.: The rate of substrate cycling between fructose 6-phosphate and fructose 1,6-bisphosphate in skeletal muscle. Biochem. J. *221:* 153–161 (1984).
7   Challiss, R.A.J.; Arch, J.R.S.; Crabtree, B.; Newsholme, E.A.: Measurement of the rate of substrate cycling between fructose 6-phosphate and fructose 1,6-bisphosphate in skeletal muscle by using a single-isotope technique. Biochem. J. *223:* 849–853 (1984).

8   Challiss, R.A.J.; Arch, J.R.S.; Newsholme, E.A.: The rate of substrate cycling between fructose 6-phosphate and fructose 1,6-bisphosphate in skeletal muscle from cold-exposed, hyperthyroid or acutely-exercised rats. Biochem. J. *231:* 217–220 (1985).

9   Cooney, G.; Taegmeyer, H.; Newsholme, E.A.: Tricarboxylic acid cycle flux and enzyme activities in the isolated working heart. Biochem. J. *200:* 701–107 (1981).

10 Crabtree, B.; Newsholme, E.A.: The activities of phosphorylase hexokinase, phosphofructokinase, lactate dehydrogenase and glycerol-3-phosphate dehydrogenase in muscle from vertebrates and invertebrates. Biochem. J. *126:* 49–58 (1972).

11 Crabtree, B.; Newsholme, E.A.: A quantitative approach to metabolic control. Curr. Top. cell. Regul. *25:* 21–76 (1985).

12 Crabtree, B.; Newsholme, E.A.: Control coefficients: their derivation and interpretation. Biochem. J. *242:* 113–120 (1987).

13 Margaria, R.: Biomechanics and energetics of muscular exercise (Clarendon Press, Oxford 1976).

14 Newsholme, E.A.; Crabtree, B.: Substrate cycle in metabolic regulation and heat generation. Biochem. Soc. Symp. *41:* 61–110 (1976).

15 Newsholme, E.A.; Crabtree, B.: Control of flux through metabolic pathways; in Hue, van der Werve, Short term control in the liver, pp. 3–18 (Elsevier/North-Holland, Amsterdam 1981).

16 Newsholme, E.A.; Leech, A.R.: Biochemistry for the medical sciences (Wiley, Chichester 1983).

17 Leech, A.R.; Beis, I.; Newsholme, E.A.: Radiochemical assays for creatine kinase and arginine kinase using rapid ion exchange separations. Analyt. Biochem. *90:* 561–575 (1978).

18 Newsholme, E.A.; Challiss, R.A.J.; Leighton, B.; Lozeman, F.J.; Budohoski, L.: A common mechanism for defective thermogenesis and insulin resistance. Nutrition *3:* 195–200 (1987).

19 Paul, J.: Comparative studies on the citric acid cycle; D. Phil. thesis, University of Oxford (1979).

20 Sutherland, E.W.; Oye, I.; Butcher, R.W.: The action of epinephrine and the role of adenyl cyclase system in hormone action. Recent Prog. Horm. Res. *21:* 623–642 (1965).

21 Booth, F.: Sports medicine (in press, 1988).

22 Wolfe, R.R.; Herndon, D.N.; Jahoor, F.; Miyoski, H.; Wolfe, M.: Effect of severe burn injury on substrate cycling by glucose and fatty acids. New Engl. J. Med. *317:* 403–408.

23 Zammit, V.A.; Newsholme, E.A.: The maximum activities of hexokinase, phosphorylase, phosphofructokinase, glycerol phosphate dehydrogenases, lactate dehydrogenase, octopine dehydrogenase, phosphoenolpyruvate carboxykinase, nucleoside diphosphatekinase, glutamate-oxaloacetate transaminase and arginine kinase in relation to carbohydrate utilization in muscles from marine invertebrates. Biochem. J. *160:* 447–462 (1976).

E.A. Newsholme, MD, Department of Biochemistry, University of Oxford,
South Parks Road, Oxford, OX1 3QU (UK)

Poortmans JR (ed): Principles of Exercise Biochemistry.
Med Sport Sci. Basel, Karger, 1988, vol 27, pp 78–119.

Carbohydrate Metabolism[1]

E. Hultman, R.C. Harris[2]

Department of Clinical Chemistry II, Karolinska Institute,
Huddinge University Hospital, Huddinge, Sweden

Introduction

For the successful completion of any physical task chemical energy
must be efficiently converted into mechanical energy at rates appropriate
to the muscles' needs. In many sporting events, e.g. tennis, rates of energy
expenditure will fluctuate greatly varying by a factor of 100 or more whilst
in others such as running, power output is more continuous. Continuous
effort will require a continuous supply of energy which itself will pose a
number of logistical problems to the body. To solve the energy problem the
body has evolved three main ways of fuelling the muscles. The first is by
phosphocreatine (PCr) which is available in only very small amounts but
can buffer against sudden increases in energy demand. At the other
extreme, fat has only a very limited ability to buffer against sudden fluctuations in demand, but is the primary endurance fuel. Carbohydrate is
utilized in all types of physical activities but is quantitatively most important during moderate to intense exercise and is the main fuel which supports the middle distance runner, the cross country skier and probably also
the disco dancer.

[1] This work was supported by grants from the Swedish Medical Research Council
(02647), the Swedish Work Health Fund (84-0261) and the Swedish Sports Confederation
(30, 44, 68/86).
[2] The authors wish to thank the entire staff at the Department of Clinical Chemistry II
for excellent collaboration in this investigation.

Carbohydrate as Provider of Energy

ATP-Producing Processes

Carbohydrate is important to muscle for the supply of chemical energy for muscle contraction. It has no other major functions in that tissue; the provision of carbon skeletons for the synthesis of amino acids and other molecules, if important, being largely a by-product of its first role.

The carbohydrates of importance to muscle are glucose and the glucose polymer, glycogen. Both are able to supply energy to the working muscle by direct rephosphorylation of ADP to ATP in 2 reactions, both of which occur in the cytoplasm:

(1) 1,3 disphosphoglycerate + ADP $\rightarrow$ ATP + 3-phosphoglycerate,

(2) phospho-enol-pyruvate + ADP $\rightarrow$ ATP + pyruvate.

Since each 6-carbon glucosyl unit will produce two 3-carbon molecules then the total yield through reactions 1 and 2 are 4 molecules of ATP. The net yield per glucosyl unit metabolized, however, is lower than this due to some ATP being used in the initial activation of the hexose molecule. For glucose this results in the loss of two molecules of ATP, one each through reactions 3 and 4:

(3) glucose + ATP $\rightarrow$ ADP + glucose 6-phosphate,

(4) fructose 6-phosphate + ATP $\rightarrow$ ADP + fructose 1,6-diphosphate,

making a net yield of 4–2 = 2 molecules of ATP. For each glucosyl unit liberated from locally stored glycogen the yield is 3 molecules, ATP being lost only through reaction 4.

The metabolism of carbohydrate is really a controlled oxidation reaction which has been divided up into a number of steps separated by manageable energy barriers. However, the eventual products of glucose or glycogen metabolism in the muscles are no different to what the products would be if direct chemical oxidation was allowed to occur. To effect a controlled oxidation, a proton (H^+) with two electrons (2e) are extracted at various points in the disassembly of the hexose unit, and combined with nicotinamide adenine nucleotide (NAD) to form reduced NAD or NADH. Totally, 2 molecules of NADH are formed during the breakdown of each glucosyl unit, whether from glucose or glycogen, to pyruvate:

(5) glyceraldehyde 3-phosphate + NAD $\rightarrow$ NADH +

1,3-diphosphoglycerate.

This occurs in the cytoplasm, but further reducing equivalents are extracted during the breakdown of pyruvate in the mitochondria, the total yield per 2 molecules of pyruvate being 8 NADH and 2 molecules of FADH (reduced flavin adenine nucleotide):

(6) pyruvate + CoA + NAD → NADH + acetyl CoA + CO_2,

(7) isocitrate + NAD → NADH + oxoglutarate,

(8) oxoglutarate + CoA + NAD → NADH + succinyl CoA + CO_2,

(9) succinate + FAD → FADH + fumerate,

(10) malate + NAD → NADH + oxaloacetate.

The NADH and FADH formed in these reactions are in turn available for oxidation via the respiratory chain of the mitochondria yielding further ATP. Per molecule of NADH the yield is 3 molecules of ATP. This makes a total of 30 from the 10 NADH produced from each glucosyl unit. The yield from FADH is slightly less being only 2 ATP per molecule, and thus adds only another 4 to the total.

Although the generation of NADH and FADH is the main route for ATP generation after the entrance of pyruvate into the tricarboxylic acid cycle a further ATP is also generated at the level of succinyl CoA. This occurs via the initial formation of GTP and accounts for a further 2 ATPs per glucosyl unit:

(11) succinyl CoA + GDP → GPT + succinate,

(12) GTP + ADP → ATP + GDP.

The final energy balance sheet for each glucosyl unit released from glycogen stored in the muscle, and oxidized fully to CO_2 and water is thus per glucosyl unit:

Anaerobic	−1 ATP (phosphorylation of fructose 6-phosphate)
yield	+4 ATP (direct cytoplasmic generation)
Aerobic	+6 ATP (mitochondrial oxidation of cytoplasmic generated NADH)
yield	+24 ATP (mitochondrial oxidation of mitochondrial generated NADH)
	+4 ATP (mitochondrial oxidation of mitochondrial generated FADH)
	+2 ATP (mitochondrial generation via GTP)
Total	39

Because of the consumption of the extra molecule of ATP, glucose will generate only 38 ATP per molecule. It has been assumed that NADH formed in the cytoplasm is capable of generating 3 ATPs via electron transport in the mitochondrion. In practice, direct transfer of cytoplasmic NADH into the mitochondria does not occur but must be mediated by one of two shuttle mechanisms. The first of these, the malate/aspartate shuttle, results in the regeneration of an equivalent amount of NADH within the mitochondria, each molecule capable of generating 3 ATPs. However, a second shuttle, which proceeds via the formation of glycerol 3-phosphate leads instead to the formation of FADH, and in this case only 2 ATPs are generated. Although this second route appears to be important in insect flight muscle, low activities of mitochondrial FAD-linked glycerol 3-phosphate dehydrogenase [42] indicate that it is of relatively minor importance in mammalian muscle. We have therefore based our calculations on the belief that the malate/aspartate shuttle is quantitatively the most important.

Rates of ATP Generation from Different Fuels

The contribution of the different fuel supplies to the total energy output by the muscle will vary both with the intensity and duration of the exercise, and will be further influenced by the fitness of the individual, the nutritional status both before and during the exercise, the level of anxiety, and even by the environment (altitude, temperature and humidity). Morphological differences between individuals in their muscle fiber make-up may also affect their use of the different fuels available. The maximum theoretical rate of utilization of a particular fuel for muscle contraction is determined by those enzymes concerned with its metabolism. In practice, further restraints imposed by the transport of substrates, particularly free fatty acids, and even the availability of cofactors will serve to limit the rate of utilization.

In table I the calculated maximal rates of high energy phosphate ($\sim$P) production from different fuel sources are presented together with the total amounts of $\sim$P available from these sources in muscle tissue. The $\sim$P generation from different fuels corresponds to the ATP production, except when part of the ATP store is broken down.

The *degradation of high energy phosphagens* gives an estimate of 2.6 mmol active phosphate ($\sim$P)$\cdot$s$^{-1}\cdot$kg^{-1}muscle. The value is determined by direct measurements in needle biopsy samples obtained from the quadriceps muscle during near maximum voluntary isometric contraction [10] or during tetanic electrical stimulation [89]. Others have suggested a figure nearer to 6 mmol $\sim$P$\cdot$s$^{-1}\cdot$kg^{-1} based on less direct measurements [44, 49, 128].

Anaerobic glycolysis gives an estimate of 1.4 mmol ATP$\cdot$s$^{-1}\cdot$kg^{-1} muscle, a value calculated from the changes in metabolites observed during near maximum isometric contraction [10, 61]. To sustain this rate of ATP

Table I. Maximum rate of $\sim$ P production from different substrates, and amounts available in muscle of a typical human

	Rate $mmol \cdot s^{-1} \cdot kg^{-1}$	Amount available $mmol \cdot kg^{-1}$ muscle
ATP, PCr → ADP, Cr	2.6	26
Glycogen → lactate	1.4	60–75, totally 240
Glycogen → CO_2	0.51–0.68	3,100
Glucose → CO_2	0.22	–
Fatty acids → CO_2	0.24	–?

production glycogen breakdown must proceed close to the V_{max} of glycogen phosphorylase. The total amount of ATP available from glycogen in the quadriceps muscle was calculated assuming 90% utilization of the normal store. This amounts to 80 mmol glucosyl units per kg muscle [73].

The amount of ATP that can be synthesized through formation of lactate is limited to 60–75 mmol·kg^{-1} muscle in the situation where the lactate is retained within the muscle. The associated rise in hydrogen ions will in this case inhibit further lactate formation or at least lower the peak rate at which this can occur. If, however, H$^+$ is continuously removed from the working muscle then theoretically the whole glycogen store could be used and would be sufficient to sustain ATP output for 2.5 min at the maximal rate indicated in table I.

The maximum rate of ATP synthesis from the complete *oxidation of carbohydrate* (0.51–0.68 mmol·s^{-1}·kg^{-1}) has been calculated assuming an oxygen utilization of 3–4 liters/min for carbohydrate oxidation in an untrained and trained individual (4 liters of oxygen is sufficient to totally oxidize 0.03 mol glucosyl units, generating 1.15 mol of ATP). The limiting factor for the maximum rate of carbohydrate oxidation is most probably mitochondrial electron transport determined by the availability of oxygen. The total amount of energy available is calculated for 80 mmol glycogen per kg muscle, a value that can be increased by exercise and diet manipulations [12].

The estimate of the maximum rate of ATP synthesis from the *oxidation of liver glycogen* by the muscle (0.22 mmol·s^{-1}·kg^{-1}) is very approximate and assumes a hepatic release of 5 mmol glucose·min^{-1} [73] of which 4 mmol·min^{-1} is made available to the working muscles (calculated in this case as 11 kg). During a short period of exercise most of this glycogen will

Table II. Maximum rate of $\sim P$ production from different substrates and amounts available in a 70-kg man; muscle mass estimated to 28 kg

	Rate mol·min^{-1}	Amount available mol
ATP, PCr → ADP, Cr	4.4	0.67
Glycogen → lactate	2.35	1.6, totally 6.7
Glycogen → CO$_2$	0.85–1.14	84
Liver glycogen → CO$_2$	0.37	19
Fatty acids → CO$_2$	0.40	4,000

be derived from the utilization of liver glycogen [79] corresponding to approximately 500 mmol glucose in the resting state following a mixed diet [103]. This would be sufficient to generate 19 mol ATP totally.

Free fatty acids (FFA) stored as triacylglycerols constitute the largest energy reserve in the body but exhibit a low rate of *oxidation*. The maximum rate of ATP resynthesis of 0.24 mmol·s^{-1}·kg^{-1} was calculated by McGilvery [97] based upon experimental results published by Pernow and Saltin [106]. The low rate indicates a limiting step located before formation of acetyl CoA, since the later steps are also used when carbohydrates are oxidized. The simultaneous utilization of blood-borne glucose and FFA will be important especially at the end of a prolonged exercise when the local carbohydrate stores in normal muscle has been depleted. Glucose can be derived from the liver or from exogenous glucose ingested during the exercise [40]. Theoretically, the maximum rate of ATP production in muscle from the combined utilization of fat and blood-borne glucose could amount to 0.46 mmol·s^{-1}·kg^{-1}. This is only 10–20% lower than the maximum rate from oxidation of muscle glycogen.

In table II the maximum rate of ATP production and total amounts available from different substrates have been recalculated for the whole working muscle mass. This is assumed to be 28 kg muscle in a 70-kg man. These values are referred to in the next section where substrate choice during exercise is discussed.

Substrate Choice in Relation to Exercise Intensity and Duration

The mechanisms by which the muscle cells regulate the use of the different fuel reserves are complex. In essence FFAs constitute the main energy substrate at rest and during very light exercise. However, delay in

Table III. Rate of $\sim$ P utilization, and total amount used during track events

Activity	Rate, $mol \cdot min^{-1}$	Amounts used, mol
Rest	0.07	
100 m sprint	2.6	0.43
400 m sprint	2.3	1.72
800 m run	2.0	3.43
1,500 m run	1.7	6.00
42,200 m marathon	1.0	150.00

Modified from Fox [52].

the mobilization of the fat reserves and the transport of FFAs will incur some utilization of carbohydrates even at these loads, at least in the initial stages. At high work loads oxidative utilization of liver and muscle glycogen are progressively more important, with anaerobic utilization of muscle glycogen and PCr being used exclusively at supramaximum work loads.

In table III a number of practical examples are given showing the rates of energy expenditure during completion of different track events. The estimates of rate and total expenditure shown have been recalculated from figures given by Fox [52]. The average rate of energy expenditure during a *100 m race* was calculated as 2.6 mol ATP·min^{-1} with a total utilization of 0.43 mol. This is just above the rate which can be sustained by anaerobic glycolysis alone (2.4 mol ATP·min^{-1}; table II) and will therefore need to be supplemented by the utilization of PCr. In practice, initial power output is probably much higher than indicated by the rate of 2.6 mol·min^{-1} and much nearer to the maximum rate from PCr utilization (4.4 mol·min^{-1}). In the closing stages of the race, however, acute acidification of the muscle will tend to lower the ATP generation rate making 2.6 the overall average. In any event PCr is an obligatory fuel for this type of exercise.

During a *400 m run* the estimated expenditure rate is 2.3 mol ATP·min^{-1} with a total utilization of 1.72 mol (table III). This could be covered by anaerobic glycolysis alone but inevitably there will be some obligatory utilization of PCr during the initial stages. There will also be a minor contribution from aerobic glycolysis which will have the further advantage of lessening the accumulation of lactate. High anaerobic power output coupled with the best that can be achieved in aerobic capacity will in this case denote the successful athlete at 400 m. As race distance is extended to *800 and 1,500 m,* the rate of energy expenditure will still necessitate the use of anaerobic glycolysis but time will allow an ever-increasing contribution from aerobic glycolysis. In the case of the 1,500 m run total ATP expenditure (6 mol) is close to the maximum available from anaerobic glycolysis.

Total ATP expenditure by the *marathon runner* (150 mol) exceeds the reserve available form aerobic glycolysis and this will necessitate the additional use of fat. However, the rate of utilization is greater than for fat alone and is only just matched by aerobic glycolysis. In practice full mobilization of the fat reserves will take up to 30 min into the race during which time up to 34% of the carbohydrate store could be utilized leaving substrate equivalent to 68 mol ATP in reserve. If during the middle part of the race fat utilization contributes a full 0.4 mol ATP·min^{-1} this will necessitate the simultaneous generation of 0.7 mol·min^{-1} from aerobic glycolysis. At this rate the remaining reserve will be emptied after a further 97 min or a total of 127 min into the race. At this point the athlete will be running on fat and any reserves he can generate through hepatic glucose release or ingestion of glucose. The athlete will have little possibility to accelerate or even maintain pace. The experienced marathon runner will attempt to monitor his performance so that total carbohydrate depletion, including amounts ingested during the run, occurs just before the end.

Conclusion

ATP, in the muscle cells, is generated mainly from metabolism of fatty acids and carbohydrates with a small reserve available in the form of PCr. Metabolism of locally stored muscle glycogen yields 3 mol ATP per glucose residue in the absence of oxygen and a further 36 mol ATP when oxidized. The yield from blood borne glucose is one ATP less. Despite the greater yield from oxidation, the maximum rate of ATP supply from anaerobic metabolism is twice that obtained from oxidation. Rates of ATP supply from carbohydrate are intermediate between those from PCr utilization and fat oxidation. Carbohydrate metabolism is the major fuel source during submaximal exercise. Following muscle glycogen depletion submaximal work may be continued by utilizing a combination of blood-borne glucose and fatty acids, but work rates will inevitably decline with total carbohydrate depletion. On its own, fat is capable of sustaining only the lowest work intensities.

Regulation of Glycolysis

Glycogen is stored in muscle in close proximity to the excitation-contraction mechanism, and the enzymes responsible for its breakdown and resynthesis. Degradation is catalyzed by two enzymes, glycogen phosphorylase (EC 2.4.1.1) and debranching enzyme (EC 2.4.1.25, EC 3.2.1.33); resynthesis by the enzyme glycogen synthetase. Both glycogen phosphorylase and synthetase exist in two forms, the interconversion in either case being regulated by a phosphorylation-dephosphorylation mechanism. In the case of phosphorylase it is the phosphorylated *a* rather than the *b* form which is considered to be physiologically active, whilst for synthetase the reverse is true with the phosphorylated D form being the inactive species and the I form being active.

Phosphorylase
Enzymatic Conversion. Phosphorylase *b* is converted to the *a* form in a reaction catalyzed by phosphorylase kinase [58]. Phosphorylase kinase itself exists in two forms; a less-active phosphorylase kinase *b* that is phosphorylated to a more active *a* form by adenosine 3′,5′-cyclic monophosphate (cAMP)-dependent protein kinase. The cAMP-dependent protein kinase is activated in response to catecholamines. Both forms of phosphorylase kinase are dependent upon Ca^{2+} for activity; however, the kinase *a* is able to activate phosphorylase at Ca^{2+} concentrations of 0.1 μmol/l close to that found in resting muscle. At this concentration phosphorylase kinase *b* is essentially inactive but its activity rises as the Ca^{2+} concentration is increased, with half maximal activity expressed at 20 μmol/l [32]. Kinase *b* is still more sensitive to Ca^{2+} when troponin or calmodulin are added to the enzyme in vitro [33], and it has been suggested that Ca^{2+}-troponin is the physiological activator of the enzyme [32, 33]. In the presence of troponin the concentration of Ca^{2+} needed for activation of kinase *b* is approximately 4 μmol/l, a concentration which is attained in contracting muscle. *Whilst troponin-Ca^{2+} activation of kinase* b *provides a link between the rate of glycogen degradation and contraction frequency, phosphorylation to kinase* a *by cAMP protein kinase by-passes the need for the contraction stimulus and provides a hormonal link.* Inactivation of phosphorylase *a* is catalyzed by a phosphoprotein phosphatase as is also phosphorylase kinase *a*.

Phosphorylase in muscle is bound to glycogen in a glycogen-protein-sarcoplasmic reticulum complex which also includes phosphorylase kinase, protein kinase, phosphorylase phosphatase as well as a number of other enzymes of glycogen metabolism [47, 98]. It has been suggested by Entman et al. [47] that the phosphorylase complex is closely associated with Ca^{2+}-Mg^{2+} ATPase responsible for the ATP-dependent accumulation of Ca^{2+} in the sarcoplasmic reticulum, the former occurring on the outer and the ATPase on the inner surface of the reticulum vesicles. This would provide a very tight link between Ca^{2+} release and phosphorylase activation.

When human or animal muscle is stimulated to contract voluntarily or by electrical stimulation an almost immediate and complete transformation of phosphorylase b to a can be observed in accordance with contraction-induced Ca^{2+} activation of phosphorylase kinase. Thereafter, however, phosphorylase a reverts back to the b form despite continued contractile activity. This is observed in human muscle during voluntary contraction as well as during electrical stimulation of human and animal muscle [27, 34, 35, 75, 110, 112].

The mechanism responsible for this reversion of phosphorylase is not known but it is suggested that the uncoupling of the b to a conversion to the contraction stimulus is due initially to the a form being released as the glycogen part of the glycogen-protein complex is hydrolyzed. A simultaneous increase in phosphorylase phosphatase activity due to release from the glycogen-protein complex would complete the reversion [35].

Allosteric Regulation. Phosphorylase b is active only in the presence of AMP and/or IMP, and inhibited by ATP and glucose 6-phosphate. Studies of the effects of intense muscular contraction have shown that IMP can increase to concentrations of several $mmol \cdot l^{-1}$. A role for IMP activation of phosphorylase b during prolonged exercise has been suggested by Aragon et al. [5]. Phosphorylase a is active even in the absence of AMP [36] and is less affected by ATP and glucose 6-phosphate at normal cell concentrations.

Substrate Regulation. The importance of substrate availability to the regulation of phosphorylase activity is still in dispute. In vitro measurements of the Michaelis constant of the enzyme for glycogen indicate a value of around 1–2 mmol glucosyl $U \cdot l^{-1}$ [21], a hundredfold lower than that in normal resting muscle. Yet, in spite of this, still higher glycogen levels in the muscle have been reported to enhance the rate of breakdown [111]. Inorganic phosphate (P_i), on the other hand, clearly plays a major

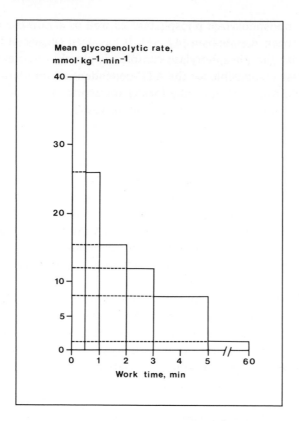

Fig. 1. Glycogenolytic rates during maximal dynamic exercise with durations from 30 s to 60 min. Work loads were chosen to produce exhaustion within the predetermined times. References are listed in the same order as the different work times [20, 57, 63, 66, 6, 13].

role in regulating activity. In the absence of AMP the K_m of phosphorylase for P_i is 26.7 mmol·l⁻¹ but in the presence of AMP is 6.8 mmol·l⁻¹ [24]. This compares to an estimated P_i concentration in resting muscle of 2–3 mmol·l⁻¹ [24, 43]. As a result of the low P_i concentration at rest, it is possible to activate the enzyme by infusion of adrenalin to obtain 90% in the *a* form and still obtain only a minimal rate of glycogen breakdown, circa 0.5–2.5 mmol glucosyl U·min⁻¹·kg⁻¹ muscle [28]. During contraction P_i is released through the breakdown of PCr increasing the availability of this substrate for phosphorylase and resulting in glycogenolytic rates as high as 30–40 mmol·min⁻¹·kg⁻¹ during short-term electrical stimulation

[75, 81, 82] and supramaximal dynamic exercise [20, 29]. P_i availability is thus an important determinant of the glycogenolytic rate and must be considered in relation to the degree of enzyme transformation induced by Ca^{2+} and/or adrcnalin.

Muscle Glycogen Degradation and Work Time. With the start of heavy exercise there is an immediate increase in glycogen degradation and anaerobic glucose utilization. The maximum rate of glycogen degradation is close to the V_{max} of phosphorylase in human quadriceps muscle (i.e. 31–50 $mmol \cdot min^{-1} \cdot kg^{-1}$) [27].

Figure 1 shows the glycogenolytic rates at different work intensities. In each case the exercise was sustained to exhaustion at an intensity calculated to be sustained for periods between 30 s and 60 min [6, 13, 20, 61, 78]. Previous studies have shown an inverse relationship between the PCr content in contracting muscle and work intensity [84], with the result that the increase in P_i content is directly related to work load. An increased P_i is predicted also at moderate work loads. In figure 2, it is shown that a work intensity sustainable for 1 h results in a glycogenolytic rate of 1–1.5 $mmol \cdot min^{-1} \cdot kg^{-1}$ muscle, which is only 3% of the maximum rate observed at an exercise sustainable for 30 s. The increase in phosphorylase a form needed to produce this rate will be very small particularly as the P_i level is increased.

The effect of the reversion of the phosphorylase transformation during continued contraction, as described above, is thus partially relieved by the increase in P_i. These two mechanisms are thus working synergetically to adjust the glycogenolytic rate to work intensity and work time. Similarly, at very high work loads increased AMP content is likely to further increase the glycogenolytic rate by increasing the affinity of both phosphorylase a and b forms for P_i [5, 27].

Hexokinase

Glycolytic units can also be derived from blood glucose taken up by working muscle. Hexokinase catalyzes the phosphorylation of glucose as it enters into the muscle cell. The reaction is accompanied by a considerable loss of free energy as heat, and is physiologically irreversible. ATP is the required phosphate source and reacts as the Mg-ATP complex. The enzyme is inhibited allosterically by the product glucose 6-phosphate.

The maximal activity in muscle is 1 $mmol$ glucose $\cdot min^{-1} \cdot kg^{-1}$ [109], sufficient to maintain an ATP production rate of 38 $mmol \cdot min^{-1} \cdot kg^{-1}$

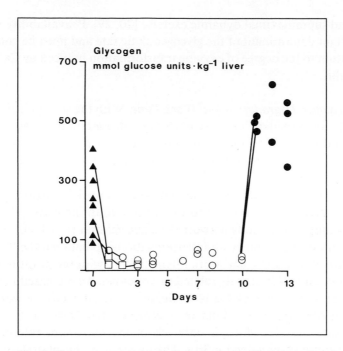

Fig. 2. Glycogen content in the liver determined in biopsy samples obtained in the postabsorbtive state after different diets in healthy subjects: after normal mixed diet (▲); after 1–3 days of total starvation (□); after 1–10 days of a carbohydrate-poor diet (○), and after 1–3 days of a carbohydrate-rich diet (●). Lines denote biopsies taken from the same subject on consecutive days. Carbohydrate-poor and carbohydrate-rich diets had the same caloric content [104].

assuming adequate glucose to be present and that all is oxidized. This figure corresponds to an oxygen uptake in a normal man of 3.7 liters·min^{-1} used exclusively for glucose oxidation in the muscle. Maximum hexokinase activity, however, can only be expressed when the glucose 6-phosphate in the cell is low and when glucose transport across the membrane is increased. Increased transport occurs when insulin is elevated and during exercise.

Phosphofructokinase
Whilst glycogen phosphorylase determines the rate of glycogen degradation it is the activity of phosphofructokinase (PFK) which dictates the overall rate of glycolytic flux to pyruvate:

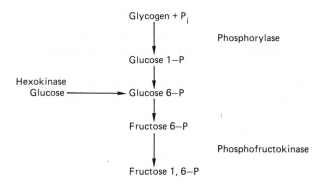

Acting as a gate to the flow of hexose units there are no other enzymes further down the chain to pyruvate which show the same highly developed control structure capable of coupling flux rate to the physiological demand for ATP. Other enzymes do exhibit some of the characteristics of regulatory enzymes, most notably pyruvate kinase and possibly also triosephosphate dehydrogenase but most probably these function only in maintaining an orderly flow through the length of the glycolytic pathway. The above PFK accumulation of hexosemonophosphates serves to balance the influx of carbohydrate either through modulation of the available P_i pool, as in the case of phosphorylase, or for hexokinase through direct inhibition by glucose 6-phosphate. Excess stimulation of phosphorylase activity by adrenalin during a controlled exercise (as opposed to the all out exercise in the true fear, flight and fight response) will result in the accumulation of still more hexose monophosphate [26] and only to a lesser degree increase glycolytic rate [89, 112]. Abolition of this response by β-blockade will lower the accumulation of hexosemonophosphates without affecting the post-PFK flux rate [25, 63]. This last study further illustrates the regulating role of PFK on glycolysis in that the flux rate was not determined simply by the availability of substrate (fructose 6-phosphate).

Recent measurements of PFK in the quadriceps muscle of trained and untrained men indicate activities of 24–35 mmol·min^{-1}·kg^{-1} measured at 25 °C [18]. This is close to the estimates of phosphorylase activity and rates of glycolysis calculated for maximally contracting muscle (table I; fig. 1).

In the transition from rest to brief tetanic contraction the rate of glycolytic flux is increased 600-fold [82]. The response of PFK to increased activity must be immediate and precisely in tune with the rate of ATP expenditure (the further production of which is after all the reason for glycolysis). This necessitates both responsive and tight metabolic control.

The essential characteristic of PFK control is allosteric inhibition by ATP [93, 125]. Although a substrate to PFK, it is also inhibitory above 5 mmol·l^{-1} by binding to a low affinity inhibitory site. Resting muscle has an ATP concentration of 7 mmol·l^{-1}. Inhibition is increased at pH below 7.2 due to pronotation of ATP binding groups at the inhibitor site [19]. The ATP inhibition is also augmented by PCr [121] with an inhibitor constant of 13 mmol·l^{-1}. This should be compared to a PCr concentration in muscle at rest of 25 mmol·l^{-1}.

Removal of ATP (and PCr) inhibition provides the primary mechanism by which PFK responds to increase in energy expenditure with the onset of contraction. This is brought about partly through the decrease in concentrations of the inhibitors and partly through increase in metabolites which either nullify the inhibitory effects of ATP, or appear to activate the enzyme directly. In approximate order of their appearance with the start of contraction these are: ADP, P$_i$, AMP, fructose 6-phosphate, glucose 1,6-bisphosphates, fructose 1,6- and 2,6-bisphosphates and under extreme conditions also NH$_4^+$. Considered individually the measured increases of each of these are usually either too small or confined to the later times of contraction to account for the increases in activity seen. NH$_4^+$, for instance, which exerts its effect in the 1- to 5-mmol·l^{-1} range [45, 122] will appear in such concentrations only after deamination of 15% or more of the adenine nucleotide pool. Termed positive modulators, it is evident from both rate considerations and detailed kinetic studies that the effects of each are synergistic with the effects of the others. However, in the rapidly changing brew of metabolites occurring in the muscle cells at this time it is impossible to say that one or other combination is dominant in modulating PFK activity or negating the inhibitory effect of ATP. The important principle is that each of the metabolites concerned with PFK regulation is responsive to increases in metabolic demand conveying information of the initiation and rate of energy expenditure and rising level of metabolic stress.

Additional control through citrate inhibition, modulating ATP inhibition, has also been described, counterbalancing ATP production from anaerobic glycolysis with that from acetyl CoA oxidation. This has the potential to coordinate fat with carbohydrate metabolism [108].

In the past, exercise physiologists have been primarily concerned with the effects of pH inhibition on PFK activity, predicting on the basis of in vitro studies, total inhibition with a pH decrease to 6.5 or there abouts. The expected result being a total loss of ATP generative capacity and onset of fatigue. However, as illustrated in a recent study of Spriet et al. [120] of the effects of repeated electrical stimulation, glycolysis may continue even when muscle pH has fallen to as low as 6.3–6.4 [83, 99]. The discrepancy between predicted and observed outcome can probably be explained on the basis of positive modulation by fructose 6-phosphate, fructose and glucose bisphosphates, ADP, AMP and NH_4^+ counteracting the pH enhancement of ATP inhibition [45], as well as a 30–40% drop in intracellular ATP concentration [83]. During prolonged contraction release of NH_4^+ from deamination of AMP, as well as the prolongation of the ADP and AMP pulse with each muscle fiber twitch, will become progressively more important in counteracting the effect of decreasing muscle pH.

Pyruvate Dehydrogenase

The decarboxylation of pyruvate to acetyl CoA is mediated by pyruvate dehydrogenase, a complex of 3 enzymes located within the inner mitochondrial membrane. The overall reaction is essentially irreversible and commits pyruvate to entry into the tricarboxylic acid cycle and total oxidation. Catalysing a flux-generating step it is not surprising that the enzyme exhibits control characteristics on a par with those of phosphorylase and phosphofructokinase.

In addition to the three constituent enzymes – pyruvate dehydrogenase, dehydrolipoamide acetyltransferase and dihydrolipoamide reductase – the complex also contains a kinase and a phosphatase catalyzing an interconversion cycle. In contrast to glycogen phosphorylase, phosphorylation of pyruvate dehydrogenase results in deactivation, whilst dephosphorylation leads to activation:

$$\underset{\text{active}}{\text{PDH}a + \text{ATP}} \underset{\text{phosphatase}}{\overset{\text{kinase}}{\rightleftarrows}} \underset{\text{inactive}}{\text{PDH}b + \text{ADP}}$$

Interconversion of the two forms is controlled by a number of allosteric effectors and influenced by hormonal action. Increased ratios of ATP/ADP, acetyl CoA/CoA and NADH/NAD activate the kinase resulting in the conversion of active pyruvate dehydrogenase *a* to inactive *b*. Pyruvate inhibits the kinase whilst an increase in Ca^{2+}, even at low concentrations, activates the phosphatase. An increase in the rate of formation of pyruvate with the initiation of contraction, coupled with an increase in Ca^{2+}, will therefore favor an increase in pyruvate dehydrogenase activity. With adequate ATP supply a high mitochondrial ATP/ADP ratio will favor a reduction in dehydrogenase activity. The inhibitory effect of an increased NADH/NAD ratio will coordinate pyruvate oxidation with electron transport chain activity, whilst the inhibitory effect of a raised acetyl CoA/CoA ratio will guard against excessive accumulation of acetyl CoA when its rate of formation outstrips its rate of condensation with oxaloacetate. Accumulation of acetyl CoA would very quickly deplete the CoA pool available in the mitochondria leading in turn to the inhibition of the TCA cycle at the level of 2-oxoglutarate dehydrogenase. This would only compound the problem since reduction in cycling would only highten the imbalance between acetyl CoA formation and removal. A transfer of acetyl units from acetyl CoA to carnitine has recently been observed during intense exercise both in man [62] and the horse [50] and likely represents a further defence against CoA depletion. Changes in the acetyl CoA/CoA ratio will also serve to coordinate pyruvate oxidation with that of fatty acids and ketone bodies [54], an increase in β-oxidation raising the acetyl CoA/CoA ratio will inhibit pyruvate dehydrogenase.

Conclusion

Glycogen utilization during exercise is initiated by Ca^{2+} release from the sarcoplasmic reticulum. The rate of breakdown and further metabolism is governed by glycogen phosphorylase, phosphofructokinase and the pyruvate dehydrogenase complex. All three enzymes exhibit complex regulatory control characteristics through which information on the contraction state, demand for energy and availability of substrates are conveyed, as well as the contribution being made by fat metabolism. Physiological control of phosphorylase is exercised through interconversion between inactive and active forms, and by the availability of P_i. Control of phosphofructokinase is mediated principally through inhibition by ATP and modification of this by a number of positive allosteric modulators, control of pyruvate dehydrogenase again involves interconversion between an inactive and active form but this is additionally influenced by a number of allosteric regulators.

Adaptation of Skeletal Muscle to Endurance Exercise

The increase in work capacity brought about by training has been intensively studied by measurements of oxygen uptake capacity and cardiovascular response. Studies of the adaptive response of skeletal muscle are, by comparison, more recent and in man have only been possible since the advent of the needle biopsy technique [9].

Oxidative Enzymes

Numerous studies have demonstrated that repeated bouts of prolonged endurance exercise can increase the muscles capacity to oxidize both pyruvate and fatty acids [8, 100]. The increase is largely brought about through increase in the number and volume of mitochondria present in the muscle fibers, resulting in marked increases in the activities of all the oxidative enzymes. In addition, capillary supply to the trained muscles is increased as well as the muscle myoglobin content [70].

As reviewed by Newsholme and Leach [102], the rate-limiting enzyme in the citric acid cycle is believed to be oxoglutarate dehydrogenase, catalyzing the formation of succinyl CoA from oxoglutarate and coenzyme A. As recently reported by Blomstrand et al. [18] oxoglutarate dehydrogenase activity in the quadriceps muscle of man was increased 39 and 90%, respectively, in medium-trained and well-trained subjects, showing increases of 21 and 49% in maximal rate of oxygen uptake. From the activities of oxoglutarate dehydrogenase estimates of the maximum rate of ATP production from oxidation were calculated and ranged from 0.22 $mmol \cdot s^{-1} \cdot kg^{-1}$ muscle in untrained female subjects to 0.48 $mmol \cdot s^{-1} \cdot kg^{-1}$ in trained male subjects. Enzyme measurements were made at 25 °C, but after adjustment to 37 °C these rates are close to those presented in table I for ATP production from aerobic metabolism and calculated from oxygen uptake capacity. At least twofold increases in the activities of other enzymes of the tricarboxylic acid cycle and the mitochondrial respiratory chain, e.g. citrate synthetase succinic dehydrogenase, cytochrome c reductase and cytochrome oxidase [67, 68, 115], have been reported with endurance training. Such changes are in the main secondary to the increase in mitochondrial content of the muscle, though a recent study of the adaptive response in rabbit muscle to chronic electrical stimulation indicates that changes may occur in the mitochondrial enzyme composition [30].

Glycolytic Enzymes

The effects of different training programs on the activities of the glycolytic enzymes are more controversial and seem to vary both with the animal studied and the fiber composition of the muscle, as well as the training program. As a whole, glycolytic enzymes tend to be decreased by *endurance training,* with the exception of hexokinase. In rat muscle increase in hexokinase activity is greatest in fibers with a high oxidation capacity [69]. In man, increased activity of hexokinase with training has been observed by Morgan et al. [100] and by Bylund et al. [22], but unchanged activity by Wallberg-Henricksson et al. [127] and by Blomstrand et al. [18]. The reason for these differences is not clear but in view of the rat studies by Holloszy [69] may possibly reflect differences in fiber composition between the muscles and individuals studied.

Phosphofructokinase activity has been reported unchanged or decreased in a series of training studies [56, 87, 116, 127]. The recent investigation by Blomstrand et al. [18] showed a 30% decrease in activity with training in the previously mentioned subjects showing a 90% increase in oxoglutarate dehydrogenase activity. The maximum rate of anaerobic glycolysis in these subjects, calculated from the rates of phosphofructokinase, was close to the maximum rates presented above. The differential changes in phosphofructokinase and oxoglutarate dehydrogenase reflect the increasing importance of ATP production from aerobic metabolism in endurance-trained athletes and decreased reliance on anaerobic ATP supply.

As with phosphofructokinase, lactate dehydrogenase has usually been observed to decrease with training [4, 7, 90]. However, it has been shown that the isoenzyme pattern may be changed with endurance training with a relative increase in the heart-type LHD-I form [4, 90, 118].

The effect of training on those enzymes immediately concerned with glycogen metabolism is equally confused. Phosphorylase activity was decreased in fast-twitch red muscles of rat by training [71], unchanged in human quadriceps femoris in the study of Morgan et al. [100], but increased in the study of Taylor [123]. In this latter study glycogen synthetase activity was also increased in the training group.

With heavy resistance training the adaptive response appears to be mainly brought about through hypertrophy of the fibers [55]. The result is an increase in the cross-sectional area of the muscle, which may be associated with a reduction in the oxidative potential per unit muscle mass as a result of 'growth dilution' of the available mitochondria. In this form of training enhancement of the muscles' ability to generate maximal power output is gained at the expense of endurance.

Lactate Production by Adapted Muscle

The trend towards increased levels of oxidative enzymes with endurance training is paralleled by an increase in VO_{2max}, increased capacity to perform strenous prolonged submaximal exercise, a lower rate of glycogen utilization, and a higher relative utilization of fat indicated by a lower respiratory quotient during exercise. Following training, lactate formation during submaximal exercise at the same relative work load (expressed as % of VO_{2max}) is decreased [65, 84, 114]. The reduction in lactate formation following training may be due either to an increased utilization of fat (vide infra) or decrease in the fraction of pyruvate converted to lactate. In the latter case increased pyruvate oxidation by the mitochondria will increase its rate of removal, but equally an increase in the mitochondrial capacity to oxidize NADH will also lower the formation of lactate. Trans-mitochondrial membrane transfer of NADH is accomplished by one of two shuttle enzyme systems, the so-called glycerol 3-phosphate and aspartate-malate shuttle. As discussed earlier the aspartate-malate shuttle is believed to be the principal transfer route in human skeletal muscle. In keeping with this, endurance training appears to have little effect on glycerol 3-phosphate dehydrogenase activity but increases the activity in muscle of aspartate-aminotransferase and malate dehydrogenase in both rat and man [69, 115].

These adaptations will decrease the rate of lactate formation at a given rate of glycolysis due to an increase in mitochondrial competition with lactate dehydrogenase for pyruvate.

Carbohydrate Utilization by Adapted Muscle

In submaximal exercise oxygen consumption is related to the work load and is the same in both the trained and untrained state. The rate of ATP production is also the same – matching the ATP utilization in the muscle. The rate of ATP formation by mitochondria is determined by the local concentrations of ADP and P_i. At onset of work ADP and P_i are increased in the muscle fibers due initially to ATP hydrolysis and PCr degradation. The increase in ADP and P_i are transferred into the mitochondria-stimulating activity, until the rate of ATP formation equals the rate of ATP utilization.

The level of ADP and P_i required to elicit the necessary increase in ATP rate will depend upon the amount of respiratory chains in function in the muscle. A higher number of respiratory chains will be able to cover the requirement of ATP production at a lower activity level per respiratory unit, and will need a smaller increase in ADP and P_i concentrations [69]. This means that the change in ATP and PCr, during an exercise at the same load, will be lower in a subject following training than before. Evidence for this was given in a study by Karlsson et al. [9], in which the same exercise was performed before and after a training period. ATP decrease was lower after 3 and 7 months of training as was also the decrease in PCr. Similarly, muscle and blood lactate concentrations were lower during the exercise following training.

As described earlier, the content of ATP, PCr, P_i, ADP, AMP and NH_3^+ function as important regulators controlling the rate of glycolysis in muscle [45, 124]. Smaller changes in these effectors in the trained muscle at a given submaximal rate of work and O_2 utilization will stimulate a lower rate of glycolysis.

An increased utilization of fat as energy substrate during submaximal exercise after training was shown by measurements of the respiratory exchange ratio [64, 65]. Increased enzyme capacity of marker enzymes of fatty acid oxidation has been shown in human muscle [39, 87]. It is further shown that the increased capacity of trained muscle to oxidize fatty acids comes primarily from an increase in the utilization of intermuscular tri-acylglycerol [72]. Increased beta-oxidation and turnover of the tricarboxylic acid cycle will increase the citric acid content in the cytoplasm resulting in increased inhibition of phosphofructokinase (see above).

In turn, this will result in the accumulation of hexosemonophosphates, principally glucose 6-phosphate which is a potent inhibitor of hexokinase and similarly affects phosphorylase *b*. The overall effect will be to lower the rate of glycolysis and decrease the uptake and utilization of blood-borne glucose. Constituting part of the 'glucose-fatty acid cycle' as first described by Randle et al. [108], it enables a trained subject to increase ATP supply from fatty acid oxidation and to decrease utilization of carbohydrate whether derived from muscle glycogen or from the liver.

Conclusion

In endurance training the muscle is adapted to a more economic utilization of fuel sources. The capacity of the tricarboxylic acid cycle and electron transport chain is increased enabling a higher rate of ATP formation in response to a lower ADP and P_i signal generated during contraction. The effect is a decrease in the glycolytic rate at a given work rate with a greater fraction being directed towards oxidation.

Coordination of the utilization of fatty acids and carbohydrate is achieved through changes in the citric acid concentration in the mitochondria affecting its concentration in the cytoplasm, and through changes in the acetyl CoA/CoA ratio. Both are increased by increased fat utilization. An increase in the acetyl CoA/CoA ratio will decrease the activity of pyruvate dehydrogenase, whilst citrate inhibits glycolysis at the level of phosphofructokinase.

Utilization of Liver Glycogen during Exercise

The liver is generally recognized as the only significant source of blood glucose in the postabsorptive state. The importance of glucose release from the liver for performance capacity was shown amongst others by Issekutz et al. [85] who found that dogs performing endurance exercise ceased to work as soon as the supply of newly released glucose from the liver became inadequate. Pronounced decrease in blood glucose during prolonged exercise, limiting performance capacity, has also been observed in man [11, 95].

Liver Glycogen Store and Glucose Production at Rest

The liver glycogen store in the postabsorptive state after a normal mixed diet was shown to be 270 mmol glucose units per kg liver (range 87–460) [79, 103]. Assuming a normal liver weight of 1.8 kg this corresponds to a total of 490 mmol glucose units. The size of the liver glycogen store, however, can be varied by the preceding diet and can range from 500 mmol·kg^{-1} (or 900 mmol totally) after a carbohydrate rich diet to as low as 12–73 mmol·kg^{-1} (or 20–120 mmol totally) after a carbohydrate-poor diet or total starvation (fig. 2) [104].

The rate of glucose release from the splanchnic area in the postabsorptive state has been found to be 0.8–1.1 mmol glucose·min^{-1}, and is just sufficient to cover the energy need of the brain and obligatory glycolytic tissues in the body [23]. Glucose released from the liver in the postabsorptive state is produced by gluconeogenesis from lactate, pyruvate, glycerol and amino acids, as well as from degradation of liver glycogen. The uptake and utilization of gluconeogenic substrates provides about 30% of the glucose released whilst 70% is derived from glycogen [104]. The rate of glycogen degradation is thus ~ 0.50 mmol glucose$\cdot$min^{-1}, sufficient to empty the glycogen store in liver during 1 day of starvation. This was confirmed by direct measurement of glycogen in liver biopsy samples (fig. 2). It was further observed that continued starvation or ingestion of a carbohydrate-poor diet kept the liver glycogen content low. Changing from a carbohydrate-free to a carbohydrate-rich diet, with the same caloric content, resulted in a rapid increase in the glycogen content and to values higher than seen after a normal mixed diet (fig. 2). Liver glycogen can also be resynthesized from intravenously given hexoses. The rate of resynthesis varies with the hexose given being 3–4 times as high when fructose is given compared to glucose [105].

Hepatic Glucose Production during Exercise

The hepatic glucose release during exercise is related to work load and exercise time. A survey of results from the literature [2, 3, 73, 126] are presented in figure 3, and show the relation between work load, exercise time and rate of hepatic glucose release. Glucose utilization during prolonged exercise is accounted for mainly by hepatic glucose release; the glucose store in the extracellular compartment corresponds to only 70 mmol glucose of which maximally 40 mmol can be used without severe effects to the central nervous system. Hepatic uptake of gluconeogenic precursors, lactate, glycerol and alanine increase during the exercise to about 2–3 times the basal level, corresponding to a maximal release of glucose of less than 1 mmol$\cdot$min^{-1}. Most of the glucose released is derived from degradation of liver glycogen. This was shown by direct analyses of the glycogen content in the liver after 1 h of heavy exercise. The glycogen content determined in liver biopsy samples had decreased from the normal mean value in the postabsorptive state of 270 mmol$\cdot$kg^{-1} to 125 mmol$\cdot$kg^{-1} after the exercise [82]. The glycogen degradation rate thus corresponded to about 4 mmol glucose units per min (liver weight 1.8 kg).

A marathon runner with an oxygen consumption of 4 liters$\cdot$min^{-1} and a glucose utilization rate of 4 mmol$\cdot$min^{-1}, would use about 0.5 liters$\cdot$min^{-1} of the oxygen uptake for glucose oxidation and produce 21.6 mol ATP (out of a total of 150 mol utilized) during a run completed within 150 min. The amount of glucose used would be 600 mmol. This amount is higher than the glycogen store in the liver in the postabsorptive state after a normal mixed diet and no more than about 100 mmol will be provided by gluconeogenesis. For this reason the marathon runner will be in danger of hypoglycemia during the final part of the run.

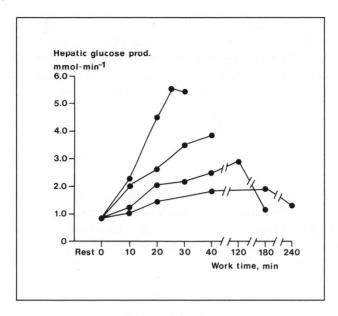

Fig. 3. Hepatic release of glucose during exercise at 4 different work loads corresponding to: 30, 50–60, 75 and 85% of VO_{2max}, respectively. The lowest glucose output was seen at a work load of 30% VO_{2max} [2] with increasing output at $\sim$ 50–60% VO_{2max} [3, 136] followed by $\sim$ 75% VO_{2max} [136] and the highest at $\sim$ 85% VO_{2max} [80].

Regulation of Hepatic Glucose Production in Exercise

Many mechanisms have been suggested for the regulation of hepatic glucose release. Traditionally, a feedback mechanism, mediated through an initial decrease in blood glucose at the start of exercise, was thought to be responsible. The decrease in glucose would stimulate the phosphorylase activity in liver directly and/or via release of hormones [101]. However, measurements of glucose in blood, in both rat and man [80, 119, 126], have shown an early increase with the initiation of exercise. The greatest increases are seen at the highest work loads [126]. At low work loads the blood glucose remains at the resting level during initiation of exercise.

Other suggested regulators of hepatic glucose release are circulating hormones (insulin, glucagon and catecholamines), as well as autonomic nerve impulses in liver tissue. A series of studies aimed at investigating these mechanisms have been performed in different experimental animals and in man but to date no particular hormonal or nervous regulator has been found showing the capacity for initiation or precise regulation of hepatic glucose release during exercise. It would appear that hepatic glyco-genolysis is started by some mechanism which is coupled to the initiation of muscle contraction and to the intensity of the exercise. Plasma insulin decrease during the exercise will enhance hepatic glucose release as will also increased plasma epinephrine and glucagon. For references see Richter et al. [113] and Galbo [53].

Liver Glycogen Utilization during Exercise after Different Diet Regimens

As shown by Nilsson and Hultman [104] the liver glycogen store is very sensitive to diet. One day of starvation or a carbohydrate poor diet decreased the liver glycogen store from 270 to 24–55 mmol·kg^{-1} liver. The glycogen content remained low during further starvation or ingestion of a carbohydrate-poor diet, whilst a carbohydrate-rich diet increased the gly-cogen content within 1 day to $\sim$ 500 mmol·kg^{-1}.

Prolonged intense exercise performed after starvation or a carbohy-drate-poor diet [11, 95] resulted in decreased endurance time and a decrease in the blood glucose content, which at least in some of the subjects was the direct cause of the inability to continue the exercise.

Measurements of hepatic glucose production during exercise after dif-ferent diets [74, 80] showed a 50% higher rate of glucose release after the carbohydrate-rich compared to the carbohydrate-poor diet (mean $\sim$ 3.5 and $\sim$ 2.4 mmol·min^{-1}, respectively). More than 50% of the glucose release after the carbohydrate-poor diet was accounted for by gluconeogen-esis from lactate, pyruvate, glycerol and alanine, whilst after the carbohy-drate-rich diet only 7% was derived from gluconeogenesis and 93% from degradation of liver glycogen. In a similar study by Björkman and Eriksson [16], a moderate 40-min exercise was performed following either a 60-hour starvation or overnight fast. Splanchnic glucose output was 0.9 mmol·min^{-1} in the 60-hour fasted subjects compared to 2.5 mmol·min^{-1} in the control group. Gluconeogenesis was responsible for 78 and 13% of hepatic glucose production, respectively.

In this last study, as well as in studies by Loy et al. [95] and Dohm et al. [46], it was shown that fat metabolism was increased during exercise following fasting. In addition, plasma insulin was lower and plasma cate-cholamine higher during exercise after fasting. All these effects will tend to decrease the utilization of blood glucose in the periphery during exercise, counteracting hypoglycemia. Nonetheless, prolonged heavy exercise is known to produce a blood glucose decrease even after a normal diet. A high carbohydrate intake the day before a prolonged heavy exercise will help in preventing exercise hypoglycemia.

Conclusion

During heavy prolonged exercise blood glucose is used at a rate of 2–4 $mmol \cdot min^{-1}$ resulting in a continuous degradation of the liver glycogen store. A normal store, range 160–800 mmol glucose units will suffice for 1–3 h of heavy exercise. A high carbohydrate intake the day before a heavy exercise bout will ensure that the liver glycogen store is filled. A carbohydrate-poor diet or starvation, on the other hand, decreases the store and can result in hypoglycemia during the exercise task, with inhibition of work performance.

Muscle Glycogen, Diet and Endurance Capacity

Diet and Muscle Glycogen

The glycogen content in muscle tissue in normally active man on a normal mixed diet is 60–115 $mmol \cdot kg^{-1}$ muscle. The content is relatively insensitive to changes in diet during periods of low physical activity. For example, during 4 days of total starvation and ordinary office work the muscle glycogen content decreased by about 40%, whilst 1 week with a carbohydrate-poor diet gave a 30% decrease. Changing to a carbohydrate-rich diet increases the glycogen store, but again the increase is only moderate when the diet is given during a period of low energy expenditure [76]. Diets given in combination with glycogen-depleting exercise have a much more dramatic effect. Resynthesis occurs also during starvation but at a low rate, while carbohydrate ingestion promotes a high resynthesis rate in the muscles which have been depleted by the exercise (fig. 4). The increase can at least initially be related to a transformation of glycogen synthetase from inactive D form to active I form in the exercised muscle [78]. The transformation is mediated by a phosphatase which is activated when glycogen is broken down.

Glycogen resynthesis in the above situation is not only fast but continues to values far above those observed in sedentary man maintained on a normal diet. In an early study from our laboratory [11], exercise was used to deplete the muscle glycogen in one leg while the glycogen content was left unchanged in the non-worked leg. Two subjects were given a carbohy-

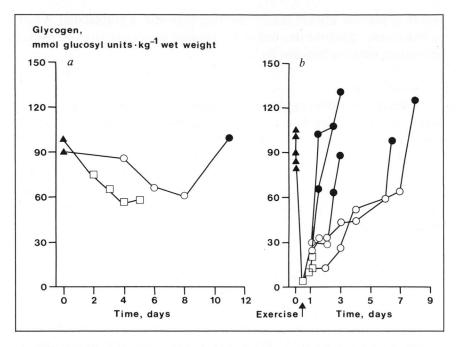

Fig. 4. a Glycogen content in quadriceps femoris muscle after a mixed diet (▲) and during 5 days of total starvation (□) in 1 subject and during a carbohydrate-poor diet (○) followed by a carbohydrate-rich diet in another subject (●) [83]. *b* Glycogen content before and after exercise (▲) followed by total starvation (□) or a carbohydrate-poor normoca-loric diet (○) and by 1–2 days of a carbohydrate-rich diet (●). Note the high rate of glycogen synthesis when a carbohydrate-rich diet is supplied compared to the low synthesis rate by a carbohydrate-poor diet [83].

drate-rich diet for 3 days. Glycogen increased in the depleted leg to 120, 193, and 230% of normal resting levels after 1, 2 and 3 days, respectively, while only a small increase occurred in the non-exercised leg (fig. 5). These studies demonstrate that glycogen resynthesis to supernormal levels is a local phenomenon restricted to the exercised, glycogen-depleted muscles.

Glycogen Stores and Exercise Capacity

The relationship between muscle glycogen content and exercise capacity was analyzed by Bergström and Hultman [13] 1967. Eight subjects performed bicycle exercise to exhaustion with a work load corresponding to ~ 80% of VO_{2max}. The exercise was done in periods of 15 min, separated

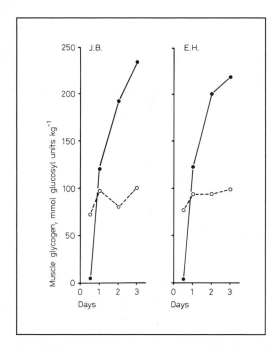

Fig. 5. The glycogen content in quadriceps femoris muscle in both the legs of 2 subjects before and after exercise performed with one leg in each subject. The exercise was followed by 3 days on a carbohydrate-rich diet. The low rate of glycogen synthesis in the rested leg (---) is contrasted to the fast increase to supernormal values in the exercised leg (——), showing the local stimulation of glycogen synthesis after glycogen depleting exercise [12].

by 15-min rest periods, during which time muscle biopsies were performed. The glycogen content in the quadriceps muscle decreased successively to values approaching zero, after ~ 70 min of exercise. At that time the predetermined work load could no longer be sustained (fig. 6).

In an early work by Christensen and Hansen [31], it was shown that a carbohydrate-rich diet during 3–7 days prior to a prolonged exercise with a submaximal work load increased the work time by 2–3 times compared to a fat diet. This observation together with the finding that muscle glycogen synthesis after exercise varied with the composition of the diet [12] and that work time at submaximal exercise seemed to be dependent on the glycogen store, initiated a study by Bergström et al. [11] aimed at deter-

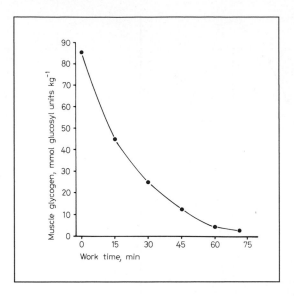

Fig. 6. The glycogen content in the vastus lateralis muscle as a function of cycling time at 80% VO_{2max} [13]. Data points are the mean values determined on eight subjects. Depletion of the glycogen store coincided with incapacity to sustain the work load.

mining a possible relationship between diet, glycogen stores and exercise capacity [11]. Muscle glycogen content was first altered by feeding different diets following a glycogen-depleting exercise, after which the subjects cycled to exhaustion at 75% VO_{2max}. A normal mixed diet was given prior to the first endurance test, a carbohydrate-poor diet before the second and a carbohydrate-rich diet prior to the third.

All three diets had the same caloric content. The carbohydrate-poor, mixed and carbohydrate-rich diets produced muscle glycogen contents of 42, 118 and 227 $mmol \cdot kg^{-1}$, respectively. The corresponding work times before exhaustion were 59, 126 and 189 min. *A close correspondence was found between initial glycogen content in the quadriceps femoris muscle and work time* (fig. 7). *Exhaustion coincided with glycogen depletion regardless of the preceding dietary regimen.* The sequence of exercise and different dietary regimens used in this study have subsequently been adapted for use in the field for muscle carbohydrate loading.

The important role of diet in optimizing initial muscle glycogen content and subsequent exercise performance has been repeatedly confirmed

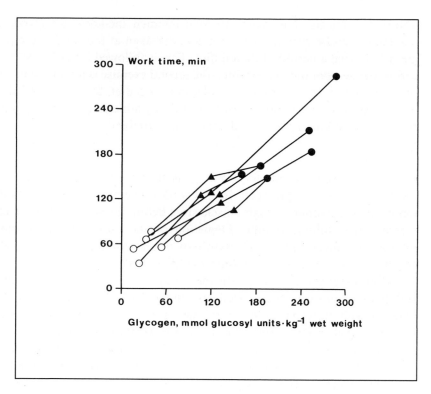

Fig. 7. The relation between initial muscle glycogen content in the quadriceps muscle and work time during bicycle exercise performed to exhaustion at a work load of 75% of VO_{2max}. The subjects worked these times, initially after a normal mixed diet, the second time after 3 days on a carbohydrate-poor diet and the third time after 3 days on a carbohydrate-rich diet [11].

[1, 37, 89, 117] and the regimens used are referred to as 'carbohydrate loading' or 'super-compensation'.

The crucial parts of the regimens are the initial depletion of the muscle glycogen stores with exhaustive exercise, followed by a high intake of carbohydrate for 3–4 days.

The 3 days of carbohydrate-poor diet used in the first experiment [11] was inserted to compare the effect of the different diets and not meant as part of a regimen aimed at carbohydrate loading. However, it has since been shown that the inclusion of 1–3 days on a carbohydrate-poor diet gives a slightly higher increase in muscle glycogen content, particularly in untrained subjects. The most effective regimen includes two glycogen-depleting exercise periods. The first is undertaken at the start of the program following a period of mixed diet. This is followed by 1–3 days on a carbohydrate-poor diet, after which the second exercise is undertaken. The program ends with 3 days on a carbohydrate-rich diet. During these 3 days no intense exercise should be done. In training athletes the carbohydrate-rich diet period after glycogen depletion is sufficient.

Conclusion

Work capacity at 60–85% VO_{2max} is limited by the availability of locally stored muscle glycogen. In this work range muscle glycogen is the main energy substrate. A high muscle store before the start of exercise can increase the endurance time. Depletion of the muscle glycogen store increases the local capacity of resynthesis resulting in an overshoot in glycogen content when a carbohydrate-rich diet is given.

High carbohydrate intake during 3 days after glycogen depletion is sufficient to optimize the glycogen store in the depleted muscles. Note that together with glycogen there is a storage of water in the muscle, about 3 liters per kg glycogen, resulting in increased weight of the muscles.

Glycogen loading of the muscles is neither necessary nor beneficial in submaximal work with shorter duration than 1 h or during single bouts of maximal or supramaximal exercise.

Utilization of Carbohydrates Taken Immediately prior to or during Exercise

The importance of dietary carbohydrate intake during the days and hours before exercise has been described – including the importance of carbohydrate-rich meals to fill both the liver and the muscle glycogen stores. An adequate diet in the days before exercise can ensure that the carbohydrate stores are optimal for energy production. Extended exercise can, however, lead to hypoglycemia if the glycogen store in the liver becomes depleted. This may lead to dysfunction of the central nervous system with dizziness and nausea, leading to inability to continue exercise. Total muscle glycogen depletion resulting in fatigue may occur during prolonged exercise at 65–85% VO_{2max}. Supplementation of carbohydrates immediately before and during prolonged exercise periods are therefore given in an attempt to avoid depletion. This is indicated both during extended periods of moderate exercise (50–70% VO_{2max}) with durations of more than 2 h and at higher work loads (70–85% VO_{2max}) with durations of more than 1 h.

Glucose Given Immediately prior to Exercise

Glucose ingestion will have two major metabolic effects: a transient increase in blood glucose level and increased insulin release. The insulin increase augments glucose uptake by muscle and adipose tissue and decreases the activity of lipase in adipose tissue. The result of a large glucose load prior to exercise is thus a decrease in the availability of FFA during the exercise, but at the same time can also result in a blood glucose decrease due to increased peripheral uptake. These effects have been demonstrated in several studies. Costill et al. [38] showed that the effect of 75 g glucose given 45 min before exercise decreased the blood glucose content and increased muscle glycogen utilization during a 30-min run at 70% VO_{2max}, compared to a control running without glucose. Similarly, 70 g of glucose given 30 min before an exhaustive exercise at 89% VO_{2max} decreased blood glucose, inhibited FFA mobilization and reduced the work time by 19% [51].

Fructose Given Immediately prior to Exercise

To avoid the insulin effect of glucose ingestion fructose has been given as an alternative. Fructose is taken up by the liver [105] and also by the muscles [13] without prior conversion to glucose by the liver. Fructose infusion increases glycogen formation in the liver at a rate ~ 3 times as high as glucose though some of the fructose is broken down to lactate and released into the blood. A fructose infusion of 1 $g \cdot h^{-1} \cdot kg^{-1}$ body weight increased blood lactate to 5–6 $mmol \cdot l^{-1}$ [15]. Pre-exercise administration of fructose was shown in two studies to have no effect on muscle glycogen utilization at 50–75% VO_{2max}, but increased total carbohydrate utilization at the expense of fat [60, 92]. In contrast, Levine et al. [94] found a lower rate of muscle glycogen utilization following fructose administration, but still total carbohydrate utilization was increased again at the expense of fat.

The ingestion of fructose before exercise has the advantage over glucose in that insulin increase with secondary risks for hypoglycemia are avoided. In addition, the increase in glycogen degradation rate seen after glucose is avoided. During extended work at loads of up to 70% VO_{2max}

plasma FFA represents a major fuel which is depressed by both glucose and fructose. The feeding of these sugars immediately prior to exercise should therefore be avoided.

Carbohydrate Ingestion during Exercise

Christensen and Hansen [31] observed that supplementation of oral glucose to exhausted man prolonged their work time. Infusion of glucose in untrained subjects during 70 min of cycling exercise at 70% VO_{2max} was shown by Bergström and Hultman [13] to reduce muscle glycogen utilization by 25%. Blood glucose increased during infusion from 4.6 to 21.5 $mmol \cdot l^{-1}$ and it was concluded that the glycogen-sparing effect was small given the drastically increased blood glucose concentration. Still, muscle glycogen is the most important fuel during exercise at this intensity.

A series of studies have since been performed with oral ingestion of glucose resulting in blood glucose increases of 2–3 $mmol \cdot l^{-1}$. In all of these [38, 41, 59, 107] glucose feeding increased overall carbohydrate utilization and decreased FFA mobilization and oxidation at both moderate and intense work loads. For carbohydrate feeding to be effective in lowering the utilization of muscle glycogen and at the same time increase endurance, the increase in exogenous glucose uptake and oxidation by muscle must provide more energy than is lost by decreased FFA oxidation. Even if the muscle glycogen utilization rate is not reduced during extended exercise, endurance may still be enhanced if the carbohydrate feeding is able to postpone liver glycogen depletion and in this way delay the onset of hypoglycemia.

Increased endurance time associated with maintenance of a normal blood glucose level has been reported in numerous studies [17, 41, 59, 86]. In contrast, no change in endurance time was found by Felig et al. [48] despite normoglycemia. Decreased rates of muscle glycogen utilization following glucose ingestion have been observed at 50 [59] and 68% VO_{2max} [17]. The glycogen utilization rate was calculated from analyses of samples taken before exercise and at exhaustion, which gives a mean rate of glycogen utilization during the whole period. A more detailed study of glycogen utilization was made by Coyle et al. [40]. They performed repeated sampling of muscle tissue for glycogen determination during the exercise periods. Seven endurance-trained cyclists exercised to fatigue twice, while ingesting a flavored placebo solution and a glucose-polymer solution (i.e. 2.0 $g \cdot kg^{-1}$ at 20 min and 0.4 $g \cdot kg^{-1}$ every 20 min thereafter). Fatigue during the placebo trial occurred after 3.02 h of exercise and was preceded by a decline in plasma glucose to 2.5 mmol·l^{-1}. When fed carbohydrate the work time was 4.02 h and the plasma glucose was maintained at the basal level. The rate of muscle glycogen utilization was the same in the two trials during the first 3 h with placebo or glucose feeding. The additional hour of exercise performed when glucose polymer was ingested was accomplished with little reliance on muscle glycogen. The R value decreased at the end of the placebo trial but was maintained when glucose was given. The authors conclude that when fed carbohydrates highly trained endurance athletes are capable of oxidizing blood glucose at relatively high rates during the late stages of prolonged strenous exercise and that this postpones fatigue. An increased utilization of blood glucose by glycogen-depleted muscle may be due to increased hexokinase activity when hexose-P inhibition of the enzyme is released as an effect of decreased glycogenolytic rate.

Conclusion

Carbohydrate intake immediately before prolonged heavy exercise is not advisable due to the risk of hypoglycemia and inhibition of FFA release from adipose tissue. On the other hand, the frequent intake of small amounts of glucose polymers during exercise counteracts blood glucose decrease and may prolong the performance time beyond the point when the muscle glycogen stores and/or blood glucose content are otherwise limiting.

Conclusions

The highest yield of ATP in contracting muscle is derived from PCr degradation and anaerobic utilization of carbohydrate. Also, when maximum force is produced by the muscle at least 50% of the ATP resynthesized is derived from glycolysis if the contraction time is more than 2–3 s. In dynamic exercise at submaximal work loads carbohydrate in the form of locally stored muscle glycogen is the most efficient source for ATP production.

Muscle glycogen degradation is initiated by Ca^{2+} release which is the trigger of muscle contraction. The rate of breakdown and the further utilization of the glycogen is regulated by contraction frequency, rate of ATP utilization and rate of ATP synthesis from other sources, governed by ATP, AMP and P_i contents, availability of glycogen and accumulation of products formed by carbohydrate and fat metabolism.

Endurance training increases the oxidative capacity of the muscles and the capacity to utilize fat, especially the muscle store of triacylglycerol. This will decrease the reliance on muscle glycogen at submaximal exercise both by increased use of fat and by a relative increase in pyruvate oxidation at the expense of lactate formation.

The liver glycogen store is utilized during submaximal exercise at a rate sufficiently high to deplete the store during prolonged work periods, resulting in hypoglycemia. The size of the glycogen store is dependent on the diet, with carbohydrate as an obligatory source for maintaining a normal glycogen level. One day of a carbohydrate-rich diet is sufficient to fill the glycogen store in the liver.

Muscle glycogen is less sensitive to diet but for resynthesis of glycogen after exercise, carbohydrate ingestion is obligatory. Glycogen depletion increases the glycogen synthesis rate, and also the capacity to store glycogen in the muscle. There is a direct relation between preexercise glycogen content and performance capacity at submaximal work loads.

Repeated intake of small amounts of carbohydrates during exercise can counteract liver glycogen depletion, and thus hypoglycemia, and prolong work time also beyond the point when the muscle glycogen store or the blood glucose content is otherwise limiting.

References

1   Ahlborg, B.; Bergström, J.; Ekelund, L.-G.; Guarnieri, G.; Harris, R.C.; Hultman, E.; Nordesjö, L.-O.: Muscle metabolism during isometric exercise performed at constant force. J. appl. Physiol. *33:* 224–228 (1972).

2   Ahlborg, G.; Felig, P.; Hagenfeldt, L.; Hendler, R.; Wahren, J.: Substrate turnover during prolonged exercise in man. Splanchnic and leg metabolism of glucose, free fatty acids, and amino acids. J. clin. Invest. *53:* 1080–1090 (1974).

3   Ahlborg, G.; Felig, P.: Lactate and glucose exchange across the forearm, legs, and splanchnic bed during and after prolonged leg exercise. J. clin. Invest. *69:* 45–54 (1982).

4   Apple, F.S.; Rogers, M.A.: Skeletal muscle lactate dehydrogenase isoenzyme alterations in men and women marathon runners. J. appl. Physiol. *61:* 477–481 (1986).

5   Aragon, J.J.; Tornheim, K.; Lowenstein, J.M.: On a possible role of IMP in the regulation of phosphorylase activity in skeletal muscle. FEBS Lett. *117:* K56–K64 (1980).

6   Åstrand, P.O.; Hultman, E.; Juhlin-Dannfelt, A.; Reynolds, G.: Disposal of lactate during and after strenous exercise in humans. J. appl. Physiol. *61:* 338–343 (1986).

7   Baldwin, K.M.; Winder, W.W.; Terjung, R.L.; Holloszy, J.O.: Glycolytic enzymes in

different types of skeletal muscle. Adaptation to exercise. Am. J. Physiol. *225:* 962–966 (1973).

8   Barnard, R.J.; Edgerton, V.R.; Peter, J.B.: Effect of exercise on skeletal muscle. I. Biochemical and histochemical properties. J. appl. Physiol. *28:* 762–766 (1970).

9   Bergström, J.: Muscle electrolytes in man. Determined by neutron activation analysis on needle biopsy specimens. A study on normal subjects, kidney patients and patients with chronic diarrhoea. Scand. J. clin. Lab. Invest. *14:* suppl. 68, pp. 1–110 (1962).

10 Bergström, J.; Harris, R.C.; Hultman, E.; Nordesjö, L.-O.: Energy rich phosphagens in dynamic and static work; in Pernow, Saltin, Muscle metabolism during exercise, pp. 341–355 (Plenum Press, New York 1971).

11 Bergström, J.; Hermansen, L.; Hultman, E.; Saltin, B.: Diet, muscle glycogen and physical performance. Acta physiol. scand. *71:* 140–150 (1967).

12 Bergström, J.; Hultman, E.: Muscle glycogen synthesis after exercise: an enhancing factor localized to the muscle cells in man. Nature, Lond. *1210:* 309–310 (1966).

13 Bergström, J.; Hultman, E.: A study of the glycogen metabolism during exercise in man. Scand. J. clin. Lab. Invest. *19:* 218–228 (1967).

14 Bergström J.; Hultman, E.: Synthesis of muscle glycogen in man after glucose and fructose infusion. Acta med. scand. *182:* 93–107 (1967).

15 Bergström, J.; Hultman, E.; Roch-Norlund, A.E.: Lactic acid accumulation in connection with fructose infusion. Acta med. scand. *184:* 359–364 (1968).

16 Björkman, O.; Eriksson, L.S.: Splanchnic glucose metabolism during leg exercise in 60-hour-fasted human subjects. Am. J. Physiol. E *245:* E443–E448 (1983).

17 Björkman, O.; Sahlin, K.; Hagenfeldt, L.; Wahren, J.: Influence of glucose and fructose ingestion on the capacity for long-term exercise in well-trained men. Clin. Physiol. *4:* 483–494 (1984).

18 Blomstrand, E.; Ekblom, B.; Newsholme, E.A.: Maximum activities of key glycolytic and oxidative enzymes in human muscle from differently trained individuals. J. Physiol. *381:* 111–118 (1986).

19 Bock, P.E.; Frieden, C.: Phosphofructokinase. I. Mechanism of pH-dependent inactivation and reactivation of the rabbit muscle enzyme. J. biol. Chem. *251:* 5630–5636 (1976).

20 Boobis, L.; Williams, C.; Wootton, S.A.: Human muscle metabolism during brief maximal exercise. J. Physiol. *338:* 21P–22P (1982).

21 Brown, D.H.; Cori, C.F.: Animal and plant polysaccharide phosphorylase; in Boyer, Lardy, Myrbäck, The enzymes, No. V (Academic Press, New York 1961).

22 Bylund, A.-C.; Bjurö, T.; Cederblad, G.; Holm, J.; Lundholm, K.; Sjöström, M.; Ängquist, K.A.; Scherstén, T.: Physical training in man. Eur. J. appl. Physiol. *36:* 151–169 (1977).

23 Cahill Jr, G.F.; Owen, O.E.: Some observations on carbohydrate metabolism in man; in Dickens, Randle, Whelan, Carbohydrate metabolism and its disorders, p. 497 (Academic Press, New York 1968).

24 Chasiotis, D.: The regulation of glycogen phosphorylase and glycogen breakdown in human skeletal muscle. Acta physiol. scand., suppl. 518, pp. 1–68 (1983).

25 Chasiotis, D.; Brandt, R.; Harris, R.C.; Hultman, E.: Effects of β-blockade on glycogen metabolism in human subjects during exercise. Am. J. Physiol. E *245:* E166–E170 (1983).

26   Chasiotis, D.; Hultman, E.: The effect of adrenalin infusion on the regulation of glycogenolysis in human muscle during isometric contraction. Acta physiol. scand. *123:* 55–60 (1985).

27   Chasiotis, D.; Sahlin, K.; Hultman, E.: Regulation of glycogenolysis in human muscle at rest and during exercise. J. appl. Physiol. *53:* 708–715 (1982).

28   Chasiotis, D.; Sahlin, K.; Hultman, E.: Regulation of glycogenolysis in human in response to epinephrine infusion. J. appl. Physiol. *54:* 45–50 (1983).

29   Cheetham, M.E.; Boobis, L.H.; Brooks, S.; Williams, C.: Human muscle metabolism during sprint running. J. appl. Physiol. *61:* 54–60 (1986).

30   Chi, M.M.-Y.; Hintz, C.S.; Henriksson, J.; Salmons, S.; Hellendahl, R.P.; Park, J.L.; Nemeth, P.M.; Lowry, O.H.: Chronic stimulation of mammalian muscle: enzyme changes in individual fibers. Am. J. Physiol. C *251:* C633–C642 (1986).

31   Christensen, E.H.; Hansen, O.: III. Arbeitsfähigkeit und Ernährung. IV. Hypoglykämie, Arbeitsfähigkeit und Ermüdung. Scand. Archs Physiol. *81:* 160–181 (1939).

32   Cohen, P.: The role of calcium ions, calmodulin and troponin in the regulation of phosphorylase kinase from rabbit skeletal muscle. Eur. J. Biochem. *111:* 563–574 (1980).

33   Cohen, P.: The role of calmodulin, troponin and cyclic AMP in the regulation of glycogen metabolism in mammalian skeletal muscle. Adv. cycl. Nucleot. Res. *14:* 345–359 (1981).

34   Conlee, R.K.; McLane, J.A.; Rennie, M.J.; Winder, W.W.; Holloszy, J.O.: Reversal of phosphorylase activation in muscle despite continued contractile activity. Am. J. Physiol. R *237:* R291–R296 (1979).

35   Constable, S.H.; Favier, R.J.; Holloszy, J.O.: Exercise and glycogen depletion: effects on ability to activate muscle phosphorylase. J. appl. Physiol. *60:* 1518–1523 (1986).

36   Cori, G.T.; Green, A.A.: Crystalline muscle phosphorylase. II. Prosthetic group. J. biol. Chem. *151:* 31–38 (1943).

37   Costill, D.L.; Miller, J.M.: Nutrition for endurance sport: carbohydrate and fluid balance. Int. J. Sports Med. *1:* 2–14 (1980).

38   Costill, D.L.; Bennett, A.; Branam, G.; Eddy, D.: Glucose ingestion at rest and during prolonged exercise. J. appl. Physiol. *34:* 764–769 (1973).

39   Costill, D.L.; Fink, W.J.; Getchell, L.H.; Ivy, J.L.; Witzmann, F.A.: Lipid metabolism in skeletal muscle of endurance-trained males and females. J. appl. Physiol. *47:* 787–791 (1979).

40   Coyle, E.F.; Coggan, A.R.; Hemmert, M.K.; Ivy, J.L.: Muscle glycogen utilization during prolonged strenous exercise when fed carbohydrate. J. appl. Physiol. *61:* 165–172 (1986).

41   Coyle, E.F.; Hadberg, J.M.; Hurley, B.F.; Martin, W.H.; Ehsani, A.A.; Holloszy, J.O.: Carbohydrate feeding during prolonged strenous exercise can delay fatigue. J. appl. Physiol. *55:* 230–235 (1983).

42   Crabtree, B.; Newsholme, E.A.: The activities of phosphorylase, hexokinase, phosphofructokinase, lactate dehydrogenase and glycerol 3-phosphate dehydrogenase in muscles from vertebrates and invertebrates. Biochem. J. *126:* 49–58 (1972).

43   Cresshull, I.; Dawson, M.J.; Edwards, R.H.T.; Gadian, D.G.; Gordon, R.E.; Radda, G.K.; Shaw, D.; Wilkie, D.R.: Human muscle analysed by ^{31}P nuclear magnetic resonance in intact subjects. Top. Magn. Reson. Spectroscop. Oxford Res. Syst. *220:* 24 (1981).

44 Davies, C.T.M.: Human power output in exercise of short duration in relation to body size and composition. Ergonomics *14:* 245–256 (1971).

45 Dobson, G.P.; Yamamoto, E.; Hochachka, P.W.: Phosphofructokinase control in muscle: nature and reversal pH-dependent ATP inhibition. Am. J. Physiol. R *250:* R71–R76 (1986).

46 Dohm, G.L.; Beeker, R.T.; Israel, R.G.; Tapscott, E.B.: Metabolic responses to exercise after fasting. J. appl. Physiol. *61:* 1363–1368 (1986).

47 Entman, M.L.; Keslensky, S.S.; Chu, A.; Van Winkle, W.B.: The sarcoplasmic reticulum-glycogenolytic complex in mammalian fast twitch skeletal muscle. J. biol. Chem. *255:* 6245–6252 (1980).

48 Felig, P.; Cherif, A.; Minagawa, A.; Wahren, J.: Hypoglycemia during prolonged exercise in normal men. New Engl. J. Med. *306:* 895–900 (1982).

49 Fletcher, J.G.; Lewis, H.E.: Photographic methods for estimating external lifting work in man. Ergonomics *2:* 114–115 (1959).

50 Foster, C.V.L.; Harris, R.C.: Changes in free carnitine in muscle of horse during high intensity exercise. Eur. J. appl. Physiol. (in press, 1987).

51 Foster, C.; Costill, D.S.; Fink, W.J.: Effects of pre-exercise feedings in endurance performance. Med. Sci. Sports *11:* 1–5 (1979).

52 Fox, E.L.: in Sports physiology; 2nd ed. CBS College Publishing (Saunders, Philadelphia 1984).

53 Galbo, H.: Endocrinology and metabolism in exercise. Curr. Probl. clin. Biochem. *11:* 26–44 (1982).

54 Garland, P.B.; Randle, P.J.: Control of pyruvate dehydrogenase in the perfused rat heart by the intracellular concentration of acetyl-coenzyme A. Biochem. J. *91:* 6C–7C (1964).

55 Goldspink, G.: Alterations in myofibril size and structure during growth, exercise, and changes in environmental temperature; in Peachy, Adrian, Geiger, Handbook of physiology: skeletal muscle, pp. 539–554 (Williams & Wilkins, Baltimore 1983).

56 Gollnick, P.D.; Armstrong, R.B.; Saubert, C.W., IV; Piehl, K.; Satlin, B.: Enzyme activity and fiber composition in skeletal muscle of untrained and trained men. J. appl. Physiol. *33:* 312–319 (1972).

57 Gollnick, P.D.; Armstrong, R.B.; Semobrowich, W.L.; Shepherd, E.E.; Saltin, B.: Glycogen depletion pattern in human skeletal muscle fibers after heavy exercise. J. appl. Physiol. *34:* 615–618 (1973).

58 Gross, S.R.; Meyer, S.E.; Regulation of phosphorylase *b* to *a* conversion in muscle. Life Sci. *14:* 404–414 (1974).

59 Hargreaves, M.; Costill, D.L.; Coggan, A.; Fink, W.J.; Nishibata, I.: Effect of carbohydrate feedings on muscle glycogen utilization and exercise performance. Med. Sci. Sports Exerc. *16:* 219–222 (1984).

60 Hargreaves, M.; Costill, D.L.; Katz, A.; Fink, W.J.: Effect of fructose ingestion on muscle glycogen usage during exercise. Med. Sci. Sports Exerc. *17:* 360–363 (1985).

61 Harris, R.C.: Muscle energy metabolism in man in response to isometric contraction. A biopsy study; thesis, University of Wales (1981).

62 Harris, R.C.; Foster, C.V.L.; Hultman, E.: Acetylcarnitine formation during intense muscular contraction in man. J. appl. Physiol. (in press, 1987).

63 Harris, R.C.; Bergström, J.; Hultman, E.: The effect of propranolol on glycogen

metabolism during exercise; in Pernow, Saltin, Muscle metabolism during exercise. Adv. exp. med. Biol., No. 11, pp. 301–305 (Plenum Press, New York 1971).

64   Henriksson, J.: Training induced adaptation of skeletal muscle and metabolism during submaximal exercise. J. Physiol. 270: 661–675 (1977).

65   Hermansen, L.; Hultman, E.; Saltin, B.: Muscle glycogen during prolonged severe exercise. Acta physiol. scand. 71: 129–139 (1967).

66   Hermansen, L.; Vaage, O.: Lactate disappearance and glycogen synthesis in human muscle after maximal exercise. Am. J. Physiol. E 233: E422–E429 (1977).

67   Holloszy, J.O.: Biochemical adaptations in muscle. Effects of exercise on mitochondrial oxygen uptake and respiratory enzyme activity in skeletal muscle. J. biol. Chem. 242: 2278–2282 (1967).

68   Holloszy, J.O.: Biochemical adaptations to exercise, aerobic metabolism; in Wilmore, Exercise and sports sciences reviews, No. I, pp. 45–71 (Academic Press, New York 1973).

69   Holloszy, J.O.: Adaptation of skeletal muscle to endurance exercise. Med. Sci. Sports 7: 155–164 (1975).

70   Holloszy, J.O.; Booth, F.W.: Biochemical adaptations to endurance exercise in muscle. A. Rev. Physiol. 38: 273–291 (1976).

71   Holloszy, J.O.; Booth, F.W.; Winder, W.W.; Fitts, R.H.: Biochemical adaptation of skeletal muscle to prolonged physical exercise; in Howald, Poortmans, Proc. 2nd Int. Symp. on Biochem. Exerc., Magglingen 1973. Metabolic adaptation to prolonged physical exercise, pp. 438–447 (Birkhäuser, Basel 1975).

72   Holloszy, J.O.; Dalsky, G.P.; Nemeth, P.M.; Hurley, B.F.; Martin, W.H., III; Hagberg, J.M.: Utilization of fat as substrate during exercise: effect of training; in Saltin, Biochemistry of exercise VI, No. 16, pp. 183–190 (Human Kinetic Publ., Champaign 1986).

73   Hultman, E.: Studies on muscle metabolism of glycogen and active phosphate in man with special reference to exercise and diet. Scand. J. clin. Lab. Invest. 19: suppl. 94, pp. 1–63 (1967).

74   Hultman, E.: Regulation of carbohydrate metabolism in the liver during rest and exercise with special reference to diet; in Landry, Orban, 3rd Int. Symp. on Biochemistry of Exercise, No. 3, pp. 99–126 (1979).

75   Hultman, E.: Carbohydrate metabolism during hard exercise and in the recovery period after exercise. Acta physiol. scand. 128: suppl. 556, pp. 75–82 (1986).

76   Hultman, E.; Bergström, J.: Muscle glycogen synthesis in relation to diet studied in normal subjects. Acta med. scand. 182: 109–117 (1967).

77   Hultman, E.; Bergström, J.; McLennan Anderson, N.: Breakdown and resynthesis of phosphorylcreatine and adenosine triphosphate in connection with muscular work. Scand. J. clin. Lab. Invest. 19: 56–66 (1967).

78   Hultman, E.; Bergström, J.; Roch-Norlund, A.E.: Glycogen storage in human skeletal muscle; in Pernow, Saltin, Muscle metabolism during exercise, pp. 273–288 (Plenum Press, New York 1971).

79   Hultman, E.; Nilsson, L.H., son: Liver glycogen in man. Effect of different diets and muscular exercise; in Pernow, Saltin, Muscle metabolism during exercise, pp. 143–141 (Plenum Press, New York 1971).

80   Hultman, E.; Nilsson, L.: Liver glycogen as a glucose-supplying source during exercise; in Keul, Int. Symp. Gravenbruch 1971, Limiting factors of physical performance, pp. 179–189 (Thieme, Stuttgart 1973).

81 Hultman, E.; Sjöholm, H.: Energy metabolism and contraction force of human skel-
    etal muscle in situ during electrical stimulation. J. Physiol. *345:* 525–532 (1983).
82 Hultman, E.; Sjöholm, H.: Substrate availability; in Knuttgen, Vogel, Poortmans,
    Int. Series on Sport Sciences, vol. 13. Biochemistry of exercise, pp. 63–75 (1983).
83 Hultman, E.; Sjöholm, H.: Biochemical causes of fatigue; in Jones, McCartney,
    McComas, Human muscle power, pp. 215–238 (Human Kinetic Publ., Champaign
    1986).
84 Hurley, B.F.; Hagberg, J.M.; Allen, W.K.; Seals, D.R.; Young, J.C.; Cuddihee, R.J.;
    Holloszy, J.O.: Effect of training on blood lactate levels during submaximal exercise.
    J. appl. Physiol. *56:* 1260–1264 (1984).
85 Issekutz, B.; Issekutz, A.C.; Nash, D.: Mobilization of energy sources in exercising
    dogs. J. appl. Physiol. *29:* 691–697 (1970).
86 Ivy, J.L.; Costill, D.L.; Fink, W.J.; Lower, R.W.: Influence of caffeine and carbohy-
    drate feedings on endurance performance. Med. Sci. Sports *11:* 6–11 (1979).
87 Jansson, E.; Kaijser, L.: Muscle adaptation to extreme endurance training in man.
    Acta physiol. scand. *100:* 315–324 (1977).
88 Jansson, E.; Hjemdahl, P.; Kaijser, L.: Epinephrine-induced changes in muscle car-
    bohydrate metabolism during exercise in male subjects. J. appl. Physiol. *60:* 1466–
    1470 (1986).
89 Karlsson, J.; Saltin, B.: Diet, muscle glycogen and endurance performance. J. appl.
    Physiol. *31:* 203–206 (1971).
90 Karlsson, J.; Frith, K.; Sjödin, B.; Gollnick, P.D.; Saltin, B.: Distribution of LDH-
    isoenzymes in human skeletal muscle. Scand. J. clin. Lab. Invest. *33:* 307–312
    (1974).
91 Karlsson, J.; Nordesjö, L.-O.; Jorfeldt, L.; Saltin, B.: Muscle lactate, ATP, and CP
    levels during exercise after physical training in man. J. appl. physiol. *33:* 199–203
    (1972).
92 Koivisto, V.A.; Härkönen, M.; Karonen, S.-L.; Groop, P.H.; Elovainio, R.; Ferran-
    nini, E.; Sacca, L.; Defronzo, R.A.: Glycogen depletion during prolonged exercise:
    influence of glucose, fructose or placebo. J. appl. Physiol. *58:* 731–737 (1985).
93 Lardy, H.A.; Parks, R.E.: Phosphofructokinase; in Gaebler, Enzymes: units of bio-
    logical structure and function, pp. 239–278 (Academic Press, New York 1956).
94 Levine, L.; Evans, W.J.; Cadarette, B.S.; Fischer, E.C.; Bullen, B.A.: Fructose and
    glucose ingestion and muscle glycogen use during submaximal exercise. J. appl.
    Physiol. *55:* 1767–1771 (1983).
95 Loy, S.F.; Conlee, R.K.; Winder, W.W.; Nelson, A.G.; Arnall, D.A.; Fischer, A.G.:
    Effects of 24-hour fast on cycling endurance time at two different intensities. J. appl.
    Physiol. *61:* 654–659 (1986).
96 Massicotte, D.; Péronnet, F.; Allah, C.; Hillaire-Marcel, C.; Ledoux, M.; Brisson, G.:
    Metabolic response to [^{13}C]glucose and [^{13}C]fructose ingestion during exercise. J.
    appl. Physiol. *61:* 1180–1184 (1986).
97 McGilvery, R.W.: The use of fuels for muscular work; in Howald, Poortmans, Met-
    abolic adaptation to prolonged physical exercise, pp. 12–30 (Birkhäuser, Basel
    1973).
98 Meyer, F.; Heilmeyer, L.M.G.; Haschke, R.H.; Fischer, E.H.: Control of phosphor-
    ylase activity in a muscle glycogen particle. I. Isolation and characterization of the
    protein-glycogen complex. J. biol. Chem. *245:* 6642–6648 (1970).

99 Meyer, R.A.; Kushmerick, M.J.; Dillon, P.F.: Intracellular pH changes in contracting fast- and slow-twitch muscles. Fed. Proc. *41:* 979 (1982).

100 Morgan, T.E.; Cobb, L.A.; Short, F.A.; Ross, R.; Gunn, D.R.: Effects of long-term
exercise on human muscle mitochondria; in Pernow, Saltin, Muscle metabolism
during exercise. Adv. exp. med. Biol., No. 11, pp. 87–95 (Plenum Press, New York
1971).

101 Newsholme, E.A.: The control of fuel utilization by muscle during exercise and
starvation. Diabetes *28:* 1–7 (1979).

102 Newsholme, E.A.; Leech, A.R.: Biochemistry for the medical sciences (Wiley, Chichester 1983).

103 Nilsson, L.H., son: Liver glycogen content in man in the postabsorptive state. Scand.
J. clin. Lab. Invest. *32:* 317–325 (1973).

104 Nilsson, L.H., son; Hultman, E.: Liver glycogen in man – the effect of total starvation or a carbohydrate-poor diet followed by carbohydrate refeedings. Scand. J. clin.
Lab. Invest. *32:* 325–330 (1973).

105 Nilsson, L.H., son; Hultman, E.: Liver and muscle glycogen in man after glucose and
fructose infusion. Scand. J. clin. Lab. Invest. *33:* 5–10 (1974).

106 Pernow, B.; Saltin, B.: Availability of substrates and capacity for prolonged heavy
exercise in man. J. appl. Physiol. *31:* 416–422 (1971).

107 Pirnay, F.; Lacroix, M.; Mosora, F.; Luyckx, A.; Lefebvre, P.: Glucose oxidation
during prolonged exercise evaluated with naturally labeled (^{13}C)glucose. J. appl.
Physiol. *43:* 258–261 (1977).

108 Randle, P.J.; Garland, P.B.; Hales, C.N.; Newsholme, E.A.: The glucose-fatty-acid
cycle. Its role in insulin sensitivity and the metabolic disturbances of diabetes mellitus. Lancet *i:* 785–789 (1963).

109 Rennie, M.J.; Edwards, R.H.T.: Carbohydrate metabolism of skeletal muscle and its
disorders; in Randle, Steiner, Whelan, Carbohydrate metabolism and its disorders,
No. 3, pp. 1–120 (Academic Press, London 1981).

110 Rennie, M.J.; Fell, R.D.; Ivy, J.L.; Holloszy, J.O.: Adrenaline reactivation of muscle
phosphorylase after deactivation during phasic contractile activity. Biosci. Rep. *2:*
323–331 (1982).

111 Richter, E.A.; Galbo, H.: Rate of glycogen breakdown and lactate release in contracting, isolated skeletal muscle is dependent upon glycogen concentration. Clin. Physiol. *5:* suppl. 4, p. 82 (1985).

112 Richter, E.A.; Ruderman, N.B.; Gavras, H.; Belur, E.R.; Galbo, H.: Muscle glycogenolysis during exercise. Dual control by epinephrine and contractions. Am. J.
Physiol. E *242:* E25–E32 (1982).

113 Richter, E.A.; Sonne, B.; Ploug, T.; Kjaer, M.; Mikines, K.; Galbo, H.: Regulation of
carbohydrate metabolism in exercise; in Saltin, Biochemistry of exercise IV. Int.
Series on Sport Sciences, No. 16, pp. 151–166 (1986).

114 Saltin, B.; Karlsson, J.: Muscle glycogen utilization during work of different intensities; in Pernow, Saltin, Muscle metabolism during exercise. Adv. exp. med. Biol.,
No. 11, pp. 289–299 (Plenum Press, New York 1971).

115 Schantz, P.G.: Plasticity of human skeletal muscle. Acta physiol. scand. *128:*
suppl. 518, pp. 1–62 (1986).

116 Schantz, P.; Billeter, R.; Henriksson, J.; Jansson, E.: Training-induced increase in
myofibrillar ATPase intermediate fibers in human skeletal muscle. Muscle Nerve *5:*
628–636 (1982).

117 Sherman, W.M.; Costill, D.L.; Fink, W.J.; Miller, J.M.: Effect of exercise-diet manipulation on muscle glycogen and its subsequent utilization during performance. Int. J. Sports Med. 2: 114–118 (1981).
118 Sjödin, B.; Thorstensson, A.; Frith, K.; Karlsson, J.: Effect of physical training on LDH activity and LDH isoenzyme pattern in human skeletal muscle. Acta physiol. scand. 97: 150–157 (1976).
119 Sonne, B.; Galbo, H.: Carbohydrate metabolism during and after exercise in rats. Studies with redioglucose. J. appl. Physiol. 59: 1627–1639 (1985).
120 Spriet, L.L.; Söderlund, K.; Bergström, M.; Hultman, E.: Skeletal muscle glycogenolysis, glycolysis, and pH during electrical stimulation in man. J. appl. Physiol. 62: 611–615 (1987).
121 Storey, K.B.; Hochachka, P.W.: Activation of muscle glycolysis: a role for creatine phosphate in phosphofructokinase regulation. FEBS Lett. 46: 337–338 (1974).
122 Sugden, P.H.; Newsholme, E.A.: The effects of ammonium, inorganic phosphate and potassium ions on the activity of phosphofructokinases from muscle and nervous tissues of vertebrates and invertebrates. Biochem. J. 150: 113–122 (1975).
123 Taylor, A.W.: The effects of exercise and training on the activities of human skeletal muscle glycogen cycle enzymes; in Howald, Poortmans, Metabolic adaptation to prolonged physical exercise, pp. 451–462 (Birkhäuser, Basel 1975).
124 Uyeda, K.; Racker, E.: Regulatory mechanisms in carbohydrate metabolism. J. biol. Chem. 240: 4682–4688 (1965).
125 Ui, M.: A role of phosphofructokinase in pH-dependent regulation of glycolysis. Biochim. biophys. Acta 124: 310–322 (1966).
126 Wahren, J.; Felig, P.; Ahlborg, G.; Jorfelt, L.: Glucose metabolism during leg exercise in man. J. clin. Invest. 50: 2715–2725 (1971).
127 Wallberg-Henriksson, H.; Gunnarsson, R.; Henriksson, J.; Östman, J.; Wahren, J.: Influence of physical training on formation of muscle capillaries in type 1 diabetes. Diabetes 33: 851–857 (1984).
128 Wilkie, D.R.: Man as a source of mechanical power. Ergonomics 3: 1–8 (1960).

Dr. Eric Hultman, Department of Clinical Chemistry II,
Huddinge University Hospital, S–141 48 Huddinge (Sweden)

Poortmans JR (ed): Principles of Exercise Biochemistry.
Med Sport Sci. Basel, Karger, 1988, vol 27, pp 120–139.

Purine Nucleotide Metabolism

K. Sahlin[a], *A. Katz*[b, 1]

[a] Department of Clinical Physiology, Karolinska Institute, Huddinge Hospital,
Huddinge, Sweden;
[b] Clinical Diabetes and Nutrition Section NIH/NIDDK, Phoenix, Ariz., USA

Introduction

Contracting muscle converts stored chemical energy to static and
kinetic energy. The immediate energy source for muscular contraction, as
for most other energy-consuming processes in the cell, is the hydrolysis of
ATP to ADP and inorganic phosphate (Pi). The concentration of ATP
within the muscle cell is, however, limited and for continuous exercise
ATP must be resynthesized at essentially the same rate as it is utilized. The
rate of ATP turnover increases almost 300 times when changing from rest
to maximal contraction and the transition can occur within a fraction of a
second. It is thus evident that the regulation of ATP resynthesis must be
very efficient.

The relative concentrations of the adenine nucleotides are regulated
by the adenylate kinase reaction (AMP + ATP $\leftrightarrow$ 2 ADP), which is considered to be close to equilibrium. Atkinson [3] termed the energy potential of
the adenine nucleotides as the energy charge (EC = (ATP + 0.5 ADP)/(ATP
+ ADP + AMP) which can theoretically be between 0 and 1. EC is not
influenced by the size of the total adenine nucleotide pool (i.e. TAN = ATP
+ ADP + AMP) and in resting muscle EC is about 0.95.

It has been known for a long time that the rate of ATP resynthesis is to
a large extent regulated by the relative concentrations of the adenine
nucleotides [36]. With the EC concept, existing knowledge was summarized into one unifying concept. Thus, it was shown by Atkinson [3] that at

[1] The authors are grateful for financial support from the Swedish Medical Research
Council (7670) and the Swedish Research Council of Sports Medicine.

a low EC metabolic sequences involved in the resynthesis of ATP were activated, whereas sequences for ATP utilization were inhibited. In contrast, at a high EC, sequences for ATP resynthesis were inhibited and sequences for utilization of ATP were stimulated. EC is, however, calculated from the total tissue contents of the adenine nucleotides, and the calculation does not discriminate between the forms that are available and unavailable for the relevant enzymatic reactions. It is further known that the adenine nucleotides are composed of different ionic forms chelated to different degrees to Mg^{2+}, K^+ and H^+ and that these forms will have different properties regarding the regulation of enzymatic activities. A change in the intracellular concentration of Mg^{2+}, K^+ or H^+ could thus affect the regulatory properties of the adenine nucleotides without affecting the calculated value for EC. Although it is clear that EC is a simplification of the metabolic regulation exerted by the adenine nucleotides it serves as a tool to unify present knowledge into a coherent theory.

It follows from the EC concept that the relative levels of the adenine nucleotides are more important for the control of energy metabolism than the absolute concentrations. The adenine nucleotide pool can decrease and the major route for this in skeletal muscle is the deamination of AMP to inosine monophosphate (IMP), which is essentially an irreversible reaction. Deamination of AMP will occur under conditions of low EC and the removal of AMP will result in an increase of EC and deamination has therefore been suggested to be of importance in maintaining a high energy potential (ATP/ADP, ATP/AMP or EC) [28].

In this brief review we will discuss how purine nucleotide metabolism is regulated in skeletal muscle and its importance for physiological processes. An in-depth review of all the aspects of this subject is not possible and the reader is referred to other reviews for additional information [3, 28, 35, 55].

Derivatives of Purine Nucleotide Catabolism

Two different routes exist for degradation of AMP in skeletal muscle: (1) deamination to IMP + ammonia (NH_3), and (2) dephosphorylation to adenosine (fig. 1). Adenosine and IMP can be further degraded to urate, which in man is excreted via the kidneys. Alternatively, IMP can be converted to adenylosuccinate (S-AMP), which can then be converted back to AMP; these three reactions constitute the purine nucleotide cycle (PNC)

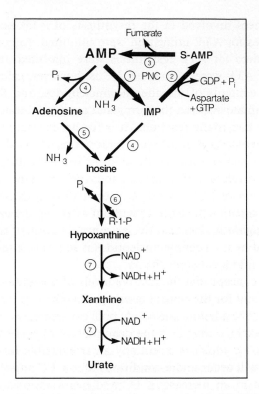

Fig. 1. Biochemical pathways associated with AMP metabolism in mammalian skeletal muscle. Enzymes catalyzing labelled reactions: (1) AMP deaminase; (2) S-AMP synthetase; (3) S-AMP lyase; (4) 5'-nucleotidase; (5) adenosine deaminase; (6) purine nucleoside phosphorylase; (7) xanthine oxidase. AMP = Adenosine monophosphate; Fum = fumarate; S-AMP, adenylsuccinate; Asp, aspartate; GT(D)P = guanosine tri(di)phosphate; R-1-P = ribose-1-phosphate; NADH and NAD = nicotinamide adenine dinucleotide − reduced and oxidized, respectively. Solid arrows denote purine nucleotide cycle [28].

[28]. NH_3 can be released into the circulation and be taken up by the liver to be used for urea production (in the urea cycle). Alternatively, NH_3 and glutamate can be converted to glutamine (see below).

The rate of NH_3 production increases during muscle contraction and is related to the contraction intensity [4, 13, 19, 31, 33]. At moderate to high contraction intensities the TAN pool decreases and there is a stoichiometric increase in IMP [24, 31, 32, 33, 45, 50, 51], which demonstrates that NH_3 is produced from the AMP deaminase reaction. The deamination of AMP occurs in both high-glycolytic and high-oxidative fast-twitch

Table I. Maximal in vitro activities of enzymes of AMP metabolism in skeletal muscle ($mmol \cdot kg^{-1}$ dry weight $\cdot min^{-1}$)

	AMP deaminase	S-AMP synthetase	S-AMP lyase	5'-Nucleo-tidase
Fast-twitch	535[a], 593[b]	–	4.26[a]	0.07[b]
Slow-twitch	142[c], 142[d]	–	4.65[c]	0.17[d]
Unidentified	1013[e]	2.97[e]	2.45[e]	–

Assays were performed on wet muscle at 30 °C. Values have been converted to dry weight (assuming a muscle H_2O content of 77%) and expresed at 35 °C (assuming a Q_{10} of 2 for all enzymes).
[a] Values are from white quadriceps of rats [62].
[b] Values are from gracilii of cats [6].
[c] Values are from solei of rats [62].
[d] Values are from solei of cats [6].
[e] Values are from unspecified rat muscle [47].

muscle provided the metabolic stress is sufficient [9, 31, 32]. In contrast, TAN depletion and IMP formation, generally do not occur in slow-twitch muscle [9, 31], but may be observed under extreme conditions (e.g. ischemia and intense stimulation) [61]. The more pronounced increase in blood NH_3 in human subjects with a high percentage of fast-twitch fibers in their muscle [8] is consistent with these animal studies.

The extent of the alternative route for AMP degradation (i.e. dephosphorylation to adenosine) will also differ between different muscles, with the slow-twitch muscle exhibiting greater increases in adenosine and inosine than the fast-twitch muscle [6, 42]. It is, however, clear that in terms of quantitatively accounting for the depletion of adenine nucleotides, the AMP deaminase reaction, resulting in IMP formation, with no further catabolism to inosine, predominates [33, 43, 46].

Part of the reason that AMP is degraded to different extents by the different routes in fast- and slow-twitch muscles may be sought in the enzymatic profiles of these muscles (table I). It can thus be calculated from the data of Bockman and McKenzie [6] that the activity of AMP deaminase is about 4 times higher in fast- than in slow-twitch muscle, respectively, and that the activity of 5-nucleotidase is about two times higher in the slow-twitch muscle (table I). Thus, under a given set of conditions, AMP degradation to IMP would be less favorable in slow- than in fast-twitch

Table II. Purine nucleotides and their catabolites in human skeletal muscle (mmol/kg dry weight)

	Rest	Exercise	
		50% $VO_{2\,max}$	100% $VO_{2\,max}$
ATP[a]	24.3±0.7	25.1±0.6	19.3±0.7
ADP[a]	3.40±0.09	3.44±0.11	3.91±0.14
AMP[a]	0.10±0.01	0.09±0.01	0.12±0.02
TAN[a]	27.8±0.8	28.7±0.5	23.3±0.7
IMP[b]	0	–	3.7 (1.6–6.4)
NH$_3$[a]	0.5±0.1	0.5±0.1	4.1±0.5
Adenosine[c]	0.025±0.005	–	0.043±0.008
Inosine[c]	0.031±0.006	–	0.44±0.10
Hypoxanthine[c]	0.04±0.01	–	0.11±0.08

Values are means ± SE.
[a] Data are from Katz et al. [24].
[b] Data are from Sahlin et al. [46].
[c] Data are from Sabina et al. [43].

muscle. An additional explanation may be that the ATP turnover rates are higher in fast-twitch than in slow-twitch muscles [7, 26]. A higher ATP turnover rate implies greater increases in the concentration of AMP in the contracting muscle [25], which will activate AMP deaminase to a greater extent (see below).

The stoichiometry between the decrease in TAN and increase in IMP, which is generally observed in contracting muscle (see above) and in vitro [29] demonstrates that the further degradation of IMP to urate is delayed. This is probably explained by the low activity of 5′-nucleotidase (table I). IMP can be converted back to AMP by the PNC but, in contrast, when IMP is dephosphorylated to xanthine and urate the purine skeleton cannot be reconverted to adenine nucleotides within the muscle. Thus, from this view-point a low activity of 5′-nucleotidase is of advantage.

The contents of the purine nucleotides and their catabolites in human skeletal muscle and the changes that occur during exercise are presented in table II. It should be noted that the values represent the total tissue contents and not the concentrations of the free forms. This point is especially critical regarding the ADP and AMP values, since it is considered that major portions of these metabolites are not available for the creatine kinase and adenylate kinase reactions, because of either protein binding or compartmentalization [58]. Since these reactions are probably close to equilibrium in vivo, the free contents of ADP and AMP have been calculated from the equilibrium constants and the tissue contents of the reactants and products in the creatine kinase and adenylate kinase reactions, respectively [30, 58]. However, in addition to considering these reactions to be at equilibrium in vivo, a number of assumptions must be made such as the free intracellular Mg^{2+} concentration. In our view, uncertainties exist regarding these calculations of the free AMP and ADP concentrations. Penetration of the details of this issue is, however, beyond the scope of this review. Future studies are required to further examine this important question.

It should also be emphasized that extremely rapid changes (in the order of milliseconds) can take place in the high energy phosphates during a contraction/relaxation cycle [16]. The actual concentration of the adenine nucleotides at the ATP utilizing sites will therefore likely be different than the total tissue contents in a muscle that is either freeze clamped in situ or frozen a few seconds after contraction [25].

Some catabolites of the adenine nucleotides have been suggested to affect specific biochemical and physiological processes. In the following the effects of IMP, NH_3 and adenosine will be discussed.

IMP has been shown to activate phosphorylase b in vitro [40] with a K_a of 5.0 mM [2]. Although under very extreme conditions the muscle concentration of IMP can approach this value [10, 21, 51], the typical IMP content that is observed or expected to occur (from the decrease in TAN) in skeletal muscle after voluntary isometric contraction (66% of maximal force) or maximal dynamic exercise to fatigue, is only about 1 mmol/l, a value well below its in vitro K_a for phosphorylase b. It should also be noted that during muscle contraction IMP accumulation does not begin until very high lactate contents are reached (50–75 mmol/kg dry weight [9, 21]. Furthermore, during recovery from intense exercise the concentration of IMP can remain high for up to 30 min or more [33, 51] although, under these conditions glycogenolysis is essentially zero. Recently, a new metabolic disorder has been discovered, where patients lack AMP deaminase in their skeletal muscle [17]. These patients cannot form any IMP but still produce lactate at approximately the same rate as control subjects during isometric contraction [49]. These observations speak against IMP, per se, having a significant effect on glycogenolysis in vivo. Thus, at present, there is little evidence that IMP plays a significant role in activating glycogenolysis in normal skeletal muscle.

Recently, it has been shown that IMP can activate glucose 1,6-bis-phosphate phosphatase (GP2-ase) of brain in vitro, with a K_a of 5 µmol/l [20]. GP2-ase is also present in skeletal muscle (5). This enzyme is responsible for the degradation of GP2, a metabolite that appears to have profound effects on key enzymes of carbohydrate metabolism [5]. GP2 has been shown to be a potent inhibitor of hexokinase [5]. During anoxia in brain [39] and diaphragm muscle [5] there are rapid and rela-tively substantial decreases in ATP and GP2. The time course and mag-nitude in the changes in these metabolites and the activity of AMP deaminase in brain [47] are consistent with an IMP activation of GP2-ase. The significance of the process may be that in an anoxic (or severely hypoxic) state the decrease in GP2 will induce an increased hexokinase activity, resulting in a greater capacity to utilize blood-borne glucose for ATP production. This would be of particular importance in tissues such as brain, which has a very low glycogen content. Whether IMP activation of GP2-ase and thus regulation of GP2 contents in contracting muscle is of significance is currently not known.

Phosphofructokinase (PFK), which is a key-point for regulation of glycolysis, is inhibited at low pH [57]. NH_3 formation has been suggested to stimulate PFK indirectly by buffering hydrogen ions, thereby raising pH and, consequently, stimulating glycolysis [28, 35]. It can, however, be cal-culated that the NH_3 accumulation during isometric and maximal dy-namic exercise to fatigue can account for the buffering of only 3% of the hydrogen ions from lactate and pyruvate [24, 25], and it therefore appears unlikely that NH_3 accumulation is of importance for PFK activation via its ability to buffer hydrogen ions.

NH_3 has also been suggested to stimulate PFK independent of its buffering effect [28, 35]. Numerous studies have shown that NH_3 increases PFK activity in vitro [52]. These studies, however, did not determine the effect of NH_3 on PFK activity in the presence of physiological concentra-tions of K^+ and inorganic phosphate (Pi). Sugden and Newsholme [52] included physiological concentrations of K^+ and Pi in their assays and demonstrated that in mammalian skeletal muscle NH_3 has no direct stim-ulating effect on PFK. Consistent with these in vitro data is the lack of agreement between the changes in muscle NH_3 concentration and the changes in lactate contents formation during exercise [9, 25], during recov-ery from exercise [24] and in AMP deaminase-deficient subjects [49]. Thus, it appears that NH_3 accumulation is of minor importance for acti-vation of glycolysis in vivo.

Adenosine has been implicated as a regulator of coronary blood flow [11, 42]. However, the significance of adenosine production and release into the circulation for regulating exercise-induced hyperemia in skeletal muscle is still controversial. Arguments favoring a major role for adenosine mediation of exercise-induced hyperemia have been put forward by several investigators [56]. On the other hand, some investigators have provided evidence that the muscle hyperemia is not mediated to a major extent by adenosine [6, 22]. These conclusions are ultimately based on the methodologies used in the various studies cited. A detailed discussion of the various techniques employed is beyond the scope of this review but our general impression is that adenosine production is of minor importance for mediating the exercise-induced hyperemia in fast-twitch muscle and that the importance in slow-twitch muscle is unclear [6, 22].

Recently, Espinal et al. [14] provided evidence that adenosine may decrease insulin sensitivity in isolated rat soleus muscle. This was demonstrated by incubating muscle in the presence of adenosine deaminase (catalyzes the degradation of adenosine) and insulin, and estimating the rates of glycogen synthesis and glycolysis. Although adenosine deaminase had no effect on the insulin-induced synthesis of glycogen, it did increase the insulin-induced glycolysis at physiological insulin concentrations. This effect was considered to be a postreceptor phenomenon and was suggested to have the same mechanism as the increased insulin sensitivity after exercise training [14]. During muscle contraction adenosine concentration increases (table II) [6] and would accordingly decrease the insulin-induced glycolysis. However, the rate of glucose utilization increases during muscle contraction and can occur in the absence of insulin [59]. It is therefore unlikely that adenosine-induced insulin resistance is of significance during exercise. Nevertheless, the possibility that adenosine-mediated insulin resistance may occur in vivo under other conditions warrants further investigations.

Regulation and AMP Deamination

It has been known for more than 50 years that contracting muscle produces NH_3 and that the major source of NH_3 in muscle is the deamination of AMP [38]. The AMP deaminase reaction is essentially irreversible and the enzyme is under complex regulatory control.

The activity of AMP deaminase is pH dependent showing an increased activity when pH decreases from 7.0 to 6.5–6.2 [48, 60]. The enzyme is at physiological concentrations of K^+ inhibited by ATP and Pi, but the inhibition can be overcome by low concentrations of ADP [41, 60]. When compared with the tissue concentration (see above) the K_m value for AMP is relatively high (0.4–1.0 mM) [41, 60] and the rate of AMP deamination will thus be proportional to the concentration of AMP. The enzyme is also regulated by GTP and GDP but these are not considered to be of importance under in vivo conditions [60].

When the kinetic characteristics of the enzyme in vitro were compared with the in vivo conditions, it was concluded that the most important regulators were AMP, ADP, inorganic phosphate and H^+ [60]. The metabolic conditions in resting muscle (high ATP; low ADP and AMP; pH = 7.0) will result in an inactive enzyme. During intense exercise the energy state of the muscle will, however, decrease (increases in ADP and AMP) and due to accumulation of lactic acid muscle pH will decrease. All of these changes will favor an activation of AMP deaminase and could thus explain the observed increase in IMP and NH_3. Currently, however, there is a controversy whether the increase in H^+ per se is of major importance for activating AMP deaminase in vivo [10, 61] or whether the effect of H^+ is an indirect effect through increases in AMP and ADP [25] (see below).

The idea that AMP deamination is primarily activated by acidosis appears to be supported by several lines of evidence. Thus, it is well documented that there is a correlation between NH_3 and lactate in blood [4, 8, 13, 24, 31, 33] as well as in muscle during exercise [19, 33]. In rat muscle stimulated in situ it was found that the calculated AMP deaminase activity increased only when the estimated muscle pH decreased to 6.6 [10]. Further, in soleus muscle where stimulation resulted in a breakdown of phosphocreatine but only a small increase in lactate, there was no accumulation of IMP [61]. When the stimulation was extended in time (10 min) both lactate and IMP accumulated in the muscle tissue [61].

It is, however, recognized that acidosis is not the sole factor responsible for activation of AMP deaminase during exercise [10, 25, 61]. Thus, when glycolysis is inhibited by iodoacetic acid there is still a pronounced depletion of adenine nucleotides [44] and formation of IMP [10, 54]. A parallel is found in vivo where patients with McArdle's disease who lack the enzyme glycogen phosphorylase and thus have no accumulation of lactate during ischemic exercise produce NH_3 at an increased rate during exercise [27, 34]. Furthermore, in a recent study where subjects performed isometric contraction to fatigue the in vivo AMP deaminase activity (calculated from the rate of TAN depletion) was not correlated to the tissue

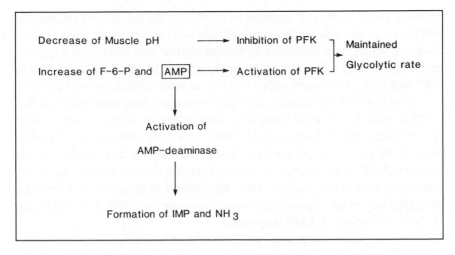

Decrease of Muscle pH   ⟶   Inhibition of PFK ⎤ Maintained

Increase of F–6–P and AMP ⟶ Activation of PFK ⎦ Glycolytic rate

Activation of

AMP–deaminase

Formation of IMP and NH$_3$

Fig. 2. Hypothesis for the regulation of AMP deaminase during muscle contraction.

lactate content but instead positively correlated to the rate of ATP turnover. These data are consistent with the idea that AMP deamination occurs during intense exercise when the capacity to rephosphorylate ADP is diminished and thus an imbalance between the rates of ATP utilization and ATP resynthesis is expected to occur (resulting in increased levels of ADP and AMP). Thus, AMP deamination may at times be associated with but is not exclusively linked to acidosis.

It is possible that the observed relationship between lactate (i.e. H^+) and NH$_3$ is not due to a direct effect of H^+ on AMP deaminase but rather is an indirect effect through an H^+-mediated inhibition of glycolysis. We have recently suggested the sequence of events described in figure 2. PFK which is one of the main regulatory enzymes in glycolysis is known to be inhibited by decreases in pH in the physiological range [57], although the inhibition can be overcome by increases in AMP and fructose 6-phosphate [57]. During sustained isometric contraction lactate accumulated and pH decreases from about 7.1 to 6.5 [45]. In spite of the decrease in pH the rate of lactate accumulation is constant during contraction [25]. Thus, any inhibition of PFK by H^+ is adequately overcome, possibly by increases in AMP and fructose 6-phosphate. The strong positive relationship between the rates of glucose 6-phosphate and fructose 6-phosphate accumulation and the glycolytic rate during isometric contraction is consistent with this explanation [23, 26]. An increase in ADP and the associated increase in AMP (due to the adenylate kinase equilibrium) will activate AMP deaminase and result in an increased formation of IMP and NH_3. If this hypothesis is correct one would expect an increase in ADP and AMP (and thus in IMP and NH_3) also under other conditions when glycolysis is impaired.

Depletion of the muscle glycogen stores by long-term exercise has recently been found to be associated with increased muscle levels of IMP [37; Broberg and Sahlin, unpubl. data] and NH_3 [Broberg and Sahlin, unpubl. data] and is yet another condition where a dissociation occurs between AMP deaminase activity and acidosis. The finding is, however, consistent with the hypothesis that impairment of glycolysis (in this case by depletion of glycogen) will evoke increases in ADP and AMP and thereby activation of AMP deaminase.

The proposed hypothesis assumes that AMP deaminase activation during muscle contraction is achieved by increases in AMP and ADP. The calculated levels of free AMP and ADP in contracting muscle were, however, considered insufficient to explain the rate of AMP deamination [10]. It should be noted, however, that the calculated values can at best correspond to the average muscle content at steady state and not to values at the ATP utilizing sites in a contracting muscle fiber, which probably are much larger (see above). Our methods to follow rapid localized changes in the cell are inadequate and therefore studies with other approaches are required to determine the relevant ADP and AMP contents.

The Purine Nucleotide Cycle and Its Functions

The reactions that constitute the PNC are presented in figure 1. It is apparent from figure 1 that one complete turn of the cycle results in the deamination of aspartate and the production of NH_3 + fumarate. Thus, fumarate and NH_3 production can occur in the absence of any net change in adenine nucleotides (see below) and could theoretically be used as an index of PNC cycling. However, NH_3 can also be formed by the glutamate dehydrogenase (GDH) reaction (glutamate + NAD^+ → α-ketoglutarate + NADH + NH_3) and be used in the glutamine synthetase reaction (NH_3 + glutamate + ATP → glutamine + ADP + Pi). The activity of GDH in skeletal muscle is, however, low or negligable [28] and deamination of AMP is the primary source of NH_3 during muscle contraction. The formation of NH_3 and glutamine in excess of the changes in TAN or IMP is a measure of the PNC cycling (i.e. both deamination of AMP and reamination of IMP back to AMP).

Meyer and Terjung [32] stimulated gastrocnemius (fast-twitch) and soleus muscle (slow-twitch) in situ to contact at high intensity. Measurements of IMP, glutamine and NH_3 showed that in gastrocnemius muscle deamination of AMP occurred during the contraction and reamination of IMP during the recovery period whereas no detectable increase in IMP occurred in soleus muscle. Thus, there was no evidence of PNC cycling during contraction in either muscle type. In a recent study in man muscle metabolites were measured before and after isometric contraction to fatigue at 66% of the maximal voluntary contraction force [25]. The increase in NH_3 was similar to the decrease in TAN and as no change in glutamine occurred it was concluded that PNC cycling did not occur. Further, during short-term bicycle exercise the decrease in TAN, production of NH_3 and removal of glutamine were measured by determination of muscle metabolites and by measuring the release of NH_3 and glutamine. During submaximal exercise (50% $VO_{2\,max}$) there was no evidence for AMP deaminase activation but during maximal exercise AMP deamination was evident since TAN was depleted by 15%. However, total NH_3 production was estimated to be essentially identical to the decrease in TAN [24] which indicates that no reamination of IMP or PNC cycling occurred.

The possibility of PNC cycling during exercise at moderate intensity has recently been investigated in man. Changes in muscle metabolites and release of NH_3 were studied during long term (about 80 min) bicycle exercise at an intensity of about 70% $VO_{2\,max}$. The data indicated that NH_3 formation was in excess of IMP formation and thus that PNC cycling probably occurred [Broberg and Sahlin, unpubl. data].

An alternative way to study PNC cycling is to block the reamination of IMP either by inhibiting S-AMP synthetase with hadacidin or by inhibiting S-AMP lyase (fig. 1) with 5-amino-4-imidazolecartoxamide ribotide (AICAR). The difference in IMP (using hadacidin) or IMP + S-AMP (using AICAR) vs. the control condition is a measure of the PNC cycling rate during the control condition. A necessary assumption is that the inhibition is selective and does not affect any other biochemical or physiological process.

Meyer and Terjung [33] used hadacidin to estimate the rates of cycling in rat muscle during exercise of varying intensities. Hadacidin had no effect on contraction force or NH_3 accumulation [33]. Their results showed no evidence for AMP deaminase activation (and thereby no cycling) during relatively mild contraction conditions. At higher contraction intensities excess IMP accumulation was observed in the hadacidin-treated rats, providing evidence that both IMP formation and reamination occurred during exercise. However, further examination of the responses of the fast-twitch red (high oxidative and glycolytic) and white (low oxidative and high glycolytic) gastrocnemius showed that IMP increased only in the white gastrocnemius and that the increase in IMP occurred during the initial phase of stimulation, while the reamination of IMP to replete the adenine nucleotide pool occurred during the latter period of stimulation. Thus, Meyer and Terjung [33] concluded that all three reactions of the PNC were not functioning concurrently in contracting muscle. However, Aragon et al. [1] also used hadacidin to study the PNC cycling in stimulated rat gastrocnemius muscle. According to their data the difference in IMP content between hadacidin-treated rats and controls increased successively during the 15 min of stimulation, which suggests concurrent deamination and reamination.

Using a similar approach, Flanagan et al. [18] used AICAR to study the rate of the PNC and to study its significance for muscle function in electrically stimulated rat gastrocnemius muscle (with intact blood supply). A higher IMP content was found in AICAR-treated rat muscles and they concluded that the PNC was operative to a significant extent. However, in contrast to the data of Meyer and Terjung [33], Flanagan et al. [18] showed that force production during moderate (aerobic) stimulation was markedly reduced in the AICAR-treated rats. Since the lactate production and the phosphocreatine degradation were greater and the tension development was lower in the AICAR-treated rats, the oxygen consumption must have been lower during AICAR treatment. It was concluded that disruption of the PNC was associated with muscle dysfunction (i.e. premature fatigue) and it was suggested that the cycle plays an anapleurotic role in providing tricarboxylic acid (TCA) cycle intermediates, via fumarate (see below), which would enhance aerobic energy production in contracting skeletal muscle. However, the contents of malate and citrate (which were used as an index of the contents of TCA cycle intermediates) were not significantly different between treatments. These data speak against the idea that AICAR reduced muscle function by decreasing the expansion of TCA cycle intermediates and it cannot be excluded that AICAR compromises muscle function by some mechanism other than disruption of the PNC.

From both the human studies and the animal experiments discussed above it appears that the PNC does not function during mild and high intensity muscle contraction. However, it appears that the PNC is active during moderate intensity contraction in human and rat fast-twitch muscle, although it is unclear whether both deamination of AMP and reamination of IMP occur concurrently to any significant extent.

The reactions of the PNC have been suggested to have the following function in contracting muscle [28]: (1) regulation of glycolysis through accumulation of NH_3; (2) regulation of glycogenolysis through accumulation of IMP; (3) expansion of the TCA cycle intermediates through production of fumarate; (4) deamination of amino acids (through aspartate) for oxidative metabolism; (5) to maintain an optimal ratio of the adenine nucleotides.

The first two possibilities have been discussed above and the last three will be discussed in the following.

One turn of the PNC results in the formation of fumarate, which could serve an anapleurotic function by providing TCA cycle intermediates during muscle contraction. The idea is that an increased concentration of TCA cycle intermediates would increase the rate of acetyl-CoA oxidation and the formation of NADH and thus increase the capacity for oxidative energy production. The TCA cycle intermediates increase during muscle contraction and about 70% of the increase is accounted for by malate and citrate [1]. Aragon et al. [1] have suggested that 70% of the increase in the TCA cycle intermediates (1.7 mmol/kg dry weight/min) during 10 min of exercise may be due to the activity of the PNC. This number is derived from studies employing hadacidin to block PNC activity. Aragon et al. [1] also attempted to account for the expansion of the TCA cycle intermediates by other mechanisms (such as the glutamate dehydrogenase, pyruvate carboxylase, malic enzyme and glutamate pyruvate transaminase (GPT) reactions). They suggested that it was unlikely that any of these reactions could account for the measured increased in the TCA cycle intermediates in their study. It should, however, be emphasized that increases in the TCA cycle intermediates (citrate + malate) also occur during high intensity exercise [15] although, as discussed above, no evidence for IMP reamination (and thus fumarate production) can be found [24]. It is thus clear that biochemical pathways other than the PNC can serve an anapleurotic function during these conditions. Glutamate-pyruvate transamination (glutamate + pyruvate → α-ketoglutarate + alanine) which results in α-ketoglutarate formation (a TCA cycle intermediate) is one possibility. The observed decrease in glutamate and increase in alanine in human muscle during low and high intensity exercise [24] as well as in contracting rat muscle [19, 32] are consistent with an anapleurotic function of this reaction.

Another proposed function of the PNC is the deamination of amino acids via aspartate, which thereby makes carbon skeletons of amino acids available for aerobic energy production. This would occur if the rate of utilization of aspartate in the PNC was matched by an equivalent rate of aspartate production via transamination. One would expect this process to be of significance when the reactions of the PNC are functioning concurrently. It is, however, noteworthy that slow-twitch muscle which to a large extent relies on oxidative metabolism does not show any PNC activity. Thus, the significance of the PNC as a process to provide energy substrates from amino acids or to increase the TCA cycle intermediates during muscle contraction is from this point of view questionable.

Deamination of AMP has also been suggested to be of importance for maintaining an optimal ATP/ADP ratio [28]. As discussed above the relative concentrations of the adenine nucleotides (i.e. ATP/ADP, ATP/AMP or EC) are of great importance for the control of energy metabolism. A decrease in EC will activate the resynthesis of ATP [3] and also the deamination of AMP (see above). Both processes will restore the energy charge or the ATP/ADP ratio to a presumed optimal value. The significance of maintaining a high ATP/ADP ratio could be due to the fact that the energy yield for the hydrolysis of ATP decreases when the concentration of the products (Pi and ADP) increases [36] and might impair ATP utilizing processes. This function finds support in that deamination of AMP occurs during contraction only when the EC is low [24, 32, 33], whereas reamination of IMP occurs during the recovery phase [24, 32]. The finding that AMP deamination occurs at high energy utilization rates in fast-twitch fibres [8, 9, 31, 33] and/or when the rate of ATP resynthesis is impaired [10, 44], is also consistent with this hypothesis.

The hypothesis that AMP deamination is necessary to maintain an optimal ATP/ADP ratio naturally leads to a discussion on muscle fatigue. It is well known that an increased deamination of AMP occurs during contraction at high intensity (see above) and further that the accumulation of IMP coincides with decreases in PCr and pH [9, 10, 33] and with a failure of the contraction process [23, 24, 45]. Contraction during ischemia will accelerate both the breakdown of PCr as well as the formation of IMP [9] and will also result in premature fatigue. Short-term low intensity exercise during aerobic conditions is usually not associated with TAN depletion or IMP formation [24, 33]. However, it has recently been shown that IMP accumulates during long-term exercise at a moderate intensity and that the accumulation coincides with the depletion of the glycogen stores and the time point of fatigue [37; Broberg and Sahlin, unpubl. observations]. Increases in IMP are thus observed at fatigue during both short-term heavy exercise and long-term moderate exercise. Muscle fatigue and IMP formation are therefore probably related to a failure of the chemical processes (PCr breakdown, glycolysis or oxidative phosphorylation) to rephosphorylate ADP at a sufficient rate resulting in increases in ADP-AMP and thereby activation of AMP deaminase and inhibition of specific ATPase systems.

Shortage of chemical fuels is also evident in other situations where carbohydrate metabolism is impaired, such as in muscles poisoned with iodoacetic acid (inhibited glycolysis), patients with McArdle's disease (impairment of glycogenolysis) or patients with a reduced activity of phosphofructokinase. In these cases the metabolic defects are also associated with a reduced contractile capacity and an increased rate of adenine nucleotide degradation [10, 27, 34]. Furthermore, patients with skeletal muscle deficient in AMP deaminase usually complain about muscle aches and weakness, and increased fatiguability. In a study by Sabina et al. [43], the capacity to perform dynamic work was found to be reduced in these patients. The increased fatiguability is consistent with the idea that contractile failure is caused by a premature increase in ADP. The finding that AMP-deficient subjects have an unaltered capacity for isometric contraction [49] does, however, speak against this hypothesis.

In conclusion, muscle fatigue is under a variety of, but not all, conditions related to an increased rate of IMP and NH_3 formation and the common link between these events is probably increases in ADP and AMP occurring locally at the ATP utilizing sites. It should, however, be emphasized that muscle fatigue is a multifactorial phenomenon and that many conditions exist where fatigue is unrelated to biochemical changes in the muscle [12].

References

1   Aragon, J.J.; Tornheim, K.; Goodman, M.N.; Lowenstein, J.M.: Replenishment of citric acid cycle intermediates by the purine nucleotide cycle in rat skeletal muscle. Curr. Top. cell. Regul. *18:* 131–149 (1981).
2   Aragon, J.J.; Tornheim, K.; Lowenstein, J.M.: On a possible role of IMP in the regulation of phosphorylase activity in skeletal muscle. Fed. Eur. biochem. Soc. Lett. *117:* suppl., pp. K56–K64 (1980).
3   Atkinson, D.E.: The energy charge of the adenylate pool as a regulatory parameter. Interaction with feedback modifiers. Biochemistry *7:* 4030–4034 (1968).

4   Babij, P.; Matthews, S.M.; Rennie, M.J.: Changes in blood ammonia lactate and amino acids in relation to workload during bicycle ergometer exercise in man. Eur. J. appl. Physiol. *50:* 405–411 (1983).

5   Beitner, R.: Glucose-1,6-bisphosphate – the regulator of carbohydrate metabolism; in Beitner, Regulation of carbohydrate metabolism, pp. 1–27 (CRC Press, Boca Raton 1985).

6   Bockman, E.; McKenzie, J.: Tissue adenosine content in active soleus and gracilis muscles of cats. Am. J. Physiol. *244:* H552–H559 (1983).

7   Crow, M.T.; Kushmeric, M.: Chemical energetics of slow and fast-twitch muscles of the mouse. J. gen. Physiol. *79:* 147–166 (1982).

8   Dudley, G.A.; Staron, R.S.; Murray, T.F.; Hagerman, R.C.; Luginbuhl, A.: Muscle fiber composition and blood ammonia levels after intense exercise in humans. J. appl. Physiol. *54:* 582–586.

9   Dudley, G.A.; Terjung, R.L.: Influence of aerobic metabolism on IMP accumulation in fast-twitch muscle. Am. J. Physiol. *248:* C37–C42 (1985).

10 Dudley, G.A.; Terjung, R.L.: Influence of acidosis on AMP deaminase activity in contracting fast-twitch muscle. Am. J. Physiol. *248:* C43–C50 (1985).

11 Edlund, A.; Fredholm, B.; Patrignani, A.; Patrono, C.; Wennmalm, Å.; Wennmalm, M.: Release of two vasodilators – adenosine and prostacyclin – from isolated rabbit hearts during controlled hypoxia. J. Physiol. *340:* 487–501 (1983).

12 Edwards, R.H.T.: Human muscle function and fatigue; in Porter, Whelan, Ciba foundation symposia Lond. 1980. Human muscle fatigue, pp. 1–18 (Pitman Medical, London 1981).

13 Eriksson, L.S.; Broberg, S.; Björkman, O.; Wahren, J.: Ammonia metabolism during exercise in man. Clin. Physiol. *5:* 325–336 (1985).

14 Espinal, J.; Chaliss, R.A.J.; Newsholme, E.A.: Effect of adenosine deaminase and adenosine analogue on insulin sensitivity in soleus muscle of the rat. FEBS Lett. *158:* 103–106 (1983).

15 Essen, B.: Studies on the regulation of metabolism in human skeletal muscle using intermittent exercise as an experimental model. Acta physiol. scand., suppl. 454, pp. 3–32 (1978).

16 Ferenczi, M.A.; Homsher, E.; Trentham, D.R.: The kinetics of magnesium adenosine triphosphate cleavage in skinned muscle fibres of the rabbit. J. Physiol. *352:* 575–599 (1984).

17 Fishbein, W.N.; Armbrustmacher, V.W.; Griffin, J.L.: Myoadenylate deaminase deficiency. A new disease of muscle. Science *341:* 534–542 (1978).

18 Flanagan, W.F.; Holmes, E.W.; Sabina, R.L.; Swain, J.L.: Importance of purine nucleotide cycle to energy production in skeletal muscle. Am. J. Physiol. *251:* C795–C802 (1986).

19 Goodman, M.N.; Lowenstein, J.M.: The purine nucleotide cycle: studies of ammonia production by skeletal muscle in situ and in perfused preparation. J. biol. Chem. *252:* 5054–5060 (1977).

20 Guha, S.K.; Rose, Z.B.: Brain glucose bisphosphate requires inosine monophosphate. J. biol. Chem. *257:* 6634–6637 (1982).

21 Harris, R.C.; Hultman, E.: Adenine nucleotide depletion in human muscle in response to intermittent stimulation in situ. J. Physiol. *365:* 78P (1985).

22 Hudlicka, O.: Regulation of muscle blood flow. Clin. Physiol. *5:* 201–229 (1985).

23   Hultman, E.; Sjöholm, H.: Energy metabolism and contraction force of human skeletal muscle in situ during electrical stimulation. J. Physiol. *345:* 525–532 (1983).

24   Katz, A.; Broberg, S.; Sahlin, K.; Wahren, J.: Muscle ammonia and amino acid metabolism during dynamic exercise in man. Clin. Physiol. *6:* 365–379 (1986).

25   Katz, A.; Sahlin, K.; Henriksson, J.: Muscle ammonia metabolism during isometric contraction in humans. Am. J. Physiol. *250:* C834–C840 (1986).

26   Katz, A.; Sahlin, K.; Henriksson, J.: Muscle ATP turnover during isometric contraction in humans. J. Appl. Physiol. *60:* 1839–1842 (1986).

27   Lewis, S.F.; Haller, R.G.: The pathophysiology of McArdle's disease: clues to regulation in exercise and fatigue. J. appl. Physiol. *61:* 391–401 (1986).

28   Lowenstein, J.M.: Ammonia production in muscle and other tissues: the purine nucleotide cycle. Physiol. Rev. *52:* 382–414 (1972).

29   Manfredi, J.P.; Holmes, E.W.: Control of the purine nucleotide cycle in extracts of rat skeletal muscle: effects of energy state and concentrations of cycle intermediates. Archs Biochem. Biophys. *233:* 515–529 (1984).

30   McGilvery, R.W.; Murray, T.W.: Calculated equilibria of phosphocreatine and adenosine phosphates during utilization of high energy phosphate by muscle. J. biol. Chem. *249:* 5845–5850 (1974).

31   Meyer, R.A.; Dudley, G.A.; Terjung, R.L.: Ammonia and IMP in different skeletal muscle fibers after exercise in rats. J. appl. Physiol. *49:* 1037–1041 (1980).

32   Meyer, R.A.; Terjung, R.A.: Differences in ammonia and adenylate metabolism in contracting fast and slow muscle. Am. J. Physiol. *237:* C111–C118 (1979).

33   Meyer, R.A.; Terjung, R.L.: AMP deamination and IMP reamination in working skeletal muscle. Am. J. Physiol. *239:* C32–C38 (1980).

34   Mineo, K.; Kono, N.; Shimizu, T.; Hara, N.; Yamada, Y.; Sumi, S.; Nonaka, K.; Tarui, S.: Excess purine degradation in exercising muscles of patients with glycogen storage disease types V and VII. J. clin. Invest. *76:* 556–560 (1985).

35   Mutch, B.J.C.; Banister, E.W.: Ammonia metabolism in exercise and fatigue: a review. Med. Sci. Sports Exer. *15:* 41–50 (1983).

36   Newsholme, E.; Start, C.: Regulation in metabolism (Wiley, London 1974).

37   Norman, B.; Jansson, E.; Sollevi, A.; Kaijser, L.: Skeletal muscle nucleotides during submaximal exercise to exhaustion. Clin. Physiol. *7:* 503–509 (1987).

38   Parnas, J.K.: Über die Ammoniakbildung im Muskel und ihren Zusammhang mit Funktion und Zustandsänderung. 6. Der Zusammenhang der Ammoniakbildung und der Umwandlung des Adeninnucleotids zu Inosinsäure. Biochem. Z. *206:* 16–38 (1929).

39   Passonneau, J.V.; Lowry, O.H.; Schulz, D.W.; Brown, J.G.: Glucose 1,6-diphosphate formation by phosphoglucomutase in mammalian tissues. J. biol. Chem. *244:* 902–909 (1969).

40   Rahim, Z.H.A.; Perrett, D.; Lutaya, G.; Griffiths, J.R.: Metabolic adaptation in phosphorylase kinase deficiency. Changes in metabolite concentrations during tetanic stimulation of mouse leg muscles. Biochem. J. *186:* 331–341 (1980).

41   Ronca-Testoni, S.; Raggi, A.; Ronca, G.: Muscle AMP aminohydrolase. III. A comparative study on the regulatory properties of skeletal muscle enzyme from various species. Biochim. biophys. Acta *198:* 101–112 (1970).

42   Rubio, R.; Berne, R.M.; Dobson, J.G.: Sites of adenosine production in cardiac and skeletal muscle. Am. J. Physiol. *225:* 938–953 (1973).

43   Sabina, R.L.; Swain, J.L.; Olanow, C.W.; Bradley, W.G.; Fishbein, W.N.; DiMauro, S.; Holmes, E.W.: Myoadenylate deaminase deficiency. Functional and metabolic abnormalities associated with disruption of the purine nucleotide cycle. J. clin. Invest. *73:* 720–730 (1984).

44   Sahlin, K.; Edström, L.; Sjöholm, L.; Hultman, E.: Effects of lactic acid accumulation and ATP decrease on muscle tension and relaxation. Am. J. Physiol. *240:* C121–C126 (1981).

45   Sahlin, K.; Harris, R.C.; Hultman, E.: Creatine kinase equilibrium and lactate content compared with muscle pH in tissue samples obtained after isometric exercise. Biochem. J. *152:* 173–180 (1975).

46   Sahlin, K.; Palmskog, G.; Hultman, E.: Adenine nucleotide and IMP contents of the quadriceps muscle in man after exercise. Pflügers Arch. *374:* 193–198 (1978).

47   Schultz, V.; Lowenstein, J.M.: Purine nucleotide cycle. Evidence for the occurrence of the cycle in brain. J. biol. Chem. *251:* 485–492 (1976).

48   Setlow, B.; Lowenstein, J.M.: Adenylate deaminase. Purification and some regulatory properties of the enzyme from calf brain. J. biol. Chem. *242:* 607–615 (1967).

49   Sinkeler, S.P.T.; Binkhorst, R.A.; Joosten, E.M.G.; Wevers, R.A.; Coerwinkel, M.M.; Oei, T.L.: AMP deaminase deficiency: study of the human skeletal muscle purine metabolism during ischaemic isometric exercise. Clin. Sci. *72:* 475–482 (1987).

50   Sjöholm, H.; Sahlin, K.; Edström, L.; Hultman, E.: Quantitative estimation of anaerobic and oxidative energy metabolism and contraction characteristics in intact human skeletal muscle during and after electrical stimulation. Clin. Physiol. *3:* 227–239 (1983).

51   Snow, D.H.; Harris, R.C.; Gash, S.D.: Metabolic response of equine muscle to intermittent maximal exercise. J. appl. Physiol. *58:* 1689–1697 (1985).

52   Sugden, P.H.; Newsholme, E.: The effects of ammonia, inorganic phosphate and potassium ions on the activity of phosphofructokinases from muscle and nervous tissues of vertebrates and invertebrates. Biochem. J. *150:* 113–122 (1975).

53   Swain, J.L.; Hines, J.J.; Sabina, R.L.; Harbury, O.L.; Holmes, E.W.: Disruption of the purine nucleotide cycle by inhibition of adenylosucciante lyase produces skeletal muscle dysfunction. J. clin. Invest. *74:* 1422–1427 (1984).

54   Terjung, R.L.; Dudley, G.A.; Meyer, R.A.: Metabolic and circulatory limitations to muscular performance at the organ level. J. exp. Biol. *115:* 307–318 (1985).

55   Terjung, R.L.; Dudley, G.A.; Meyer, R.A.; Hood, D.A.; Gorski, J.: Purine nucleotide cycle function in contracting muscle, in Saltin, 6th Int. Symp. Biochem. Exerc., Copenhagen 1985. Biochemistry of exercise. Int. Series on Sport Sciences, vol. 16, pp. 131–147 (Human Kinetics, Champaign 1986).

56   Tominaga, S.; Suzuki, T.; Nakamura, T.: Evaluation of roles of potassium, inorganic phosphate, osmolarity, pH, pCO_2 and adenosine or AMP in exercise and reactive hyperaemia in canine hindlimb muscles. Tohoku J. exp. Med. *109:* 347–363 (1973).

57   Trivedi, B.; Danforth, W.H.: Effect of pH on the kinetics of frog muscle phosphofructokinase. J. biol. Chem. *241:* 4110–4114 (1966).

58   Veech, R.L.; Lawson, J.W.R.; Cornell, N.W.; Krebs, H.A.: Cytosolic phosphorylation potential. J. biol. Chem. *254:* 6538–6547 (1979).

59 Wallberg-Henriksson, H.: Glucose transport into skeletal muscle. Influence of contractile activity, insulin catecholamines and diabetes mellitus. Acta physiol. scand., suppl. 564, pp. 1–80 (1987).

60 Wheeler, T.J.; Lowenstein, J.M.: Adenylate deaminase from rat muscle. Regulation by purine nucleotides and orthophosphate in the presence of 150 mM KCl. J. biol. Chem. 254: 8894–8999 (1979).

61 Whitlock, D.M.; Terjung, R.L.: ATP depletion in slow-twitch rat muscle of rat. Am. J. Physiol. 253: C426–C432 (1987).

62 Winder, W.W.; Terjung, R.L.; Baldwin, K.M.; Holloszy, J.O.: Effect of exercise on AMP deaminase and adenylosuccinase in rat skeletal muscle. Am. J. Physiol. 227: 1411–1414 (1974).

Dr. K. Sahlin, Department of Clinical Physiology, Huddinge Hospital,
S–141 86 Huddinge (Sweden)

Poortmans JR (ed): Principles of Exercise Biochemistry.
Med Sport Sci. Basel, Karger, 1988, vol 27, pp 140–163.

Lipid Mobilization and Utilization

Jens Bülow

Department of Clinical Physiology/Nuclear Medicine, Bispebjerg Hospital,
Copenhagen, Denmark

Introduction

Free fatty acids (FFA) have, since the demonstration of their rapid
turnover in the plasma pool by Gordon and Cherkes [50] and Dole [35],
been recognized as one of the main energy sources during rest as well as
during exercise. FFA for muscle metabolism may be derived from various
sources, primarily adipose tissue, but also from circulating lipoproteins
and from triacylglycerols stored intramuscularly while triacylglycerols
stored in the liver are not mobilized or utilized during exercise.

Since the discovery of the importance of fatty acids (FA) as an energy
source, adipose tissue has been investigated intensely and several reviews
have been given concerning the nervous, hormonal and pharmacological
regulation of adipose tissue metabolism [5, 54, 57, 85, 98].

The present review will focus mainly on the lipid metabolism during
exercise and the effect of training.

Selection of Fuel during Prolonged Exercise

Adipose tissue represents the largest energy store of the body. In nor-
mal-weight man adipose tissue contains about 400 MJ in contrast to the
energy stored as carbohydrate in liver and muscle which is about 4–5 MJ.
While the lipid and carbohydrate energy stores can undergo sizeable
changes during mobilization, the energy stored as protein, about 100 MJ,
cannot be mobilized to any greater extent without seriously affecting the
organism.

Table I. Fractional distribution of lipid and carbohydrate combustion during prolonged exercise of moderate intensity in man [upper panel modified after Bülow, 16] and dog [lower panel modified after Paul, 90]: the oxidation rates of carbohydrate and lipid in mmol/min are given for a man with a maximal oxygen uptake of 4 liters/min at a work load of 50% of VO_{2max}

	Rest	1 h	2 h	3 h	4 h
Man					
Carbohydrate, %	54	27	20	17	13
Lipid, %	46	73	80	83	87
r	0.86	0.78	0.76	0.75	0.74
Carbohydrate, mmol/min	7.2	3.6	2.6	2.3	1.7
Lipid, mmol/min	0.45	0.71	0.78	0.81	0.87
Dog					
Muscle carbohydrate, %	48	50	25	11	9
Plasma glucose, %	6	6	6	6	6
Muscle fat, %	36	20	16	13	10
Extramuscular fat, %	10	24	53	70	75
r	0.80	0.75	0.73	0.72	0.71

During heavy exercise (i.e. oxygen uptake > 70% of maximal oxygen uptake) carbohydrate is almost exclusively metabolized. In contrast, during prolonged exercise of moderate intensity there is a progressive change from carbohydrate metabolism to lipid metabolism. Initially, carbohydrate covers about 80% of the oxidative metabolism while late in exercise lipid covers up to 90%. Table I shows the fractions of oxidative metabolism covered of carbohydrate and lipid during prolonged exercise of moderate intensity, i.e. about 50% of maximal oxygen uptake.

Regulation of Lipid Metabolism during Rest

Adipose Tissue

The rate-controlling enzyme of adipose tissue lipolysis is the hormone-sensitive lipase (HSL) which catalyses the degradation of triacylglycerol to diacylglycerol and monoacylglycerol while the hydrolysis of the third FA binding is catalyzed by a specific monoacylglycerol lipase. Figure 1 shows the lipolytic pathway in adipose tissue.

In human adipose tissue noradrenaline and adrenaline are the principal lipolytic hormones. Human adipocytes possess stimulatory β-adreno-

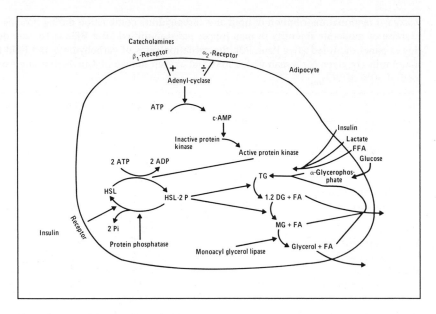

Fig. 1. Principal regulation of lipolysis in adipose tissue.

ceptors and inhibitory α-adrenoceptors [65, 66]. The activation of the hormone-sensitive lipase is effected through reversible phosphorylation catalyzed by c-AMP [5].

The only antilipolytic hormone of physiological importance in human adipocytes is insulin. The antilipolytic effect of insulin has been shown to be due to a conversion of the hormone-sensitive lipase from the phosphorylated to the dephosphorylated state [81]. This inhibitory effect of insulin is probably elicited secondary to decreased c-AMP concentration due to an activating effect on phosphodiesterase and possibly an inhibition of adenylate cyclase leading to a decrease in c-AMP-dependent phosphokinase activity [5].

Lipoproteins

Lipolysis of circulating lipoproteins is catalyzed by lipoprotein lipase (LPL). This enzyme is widely distributed in the body. The highest concentrations are found in the heart, adipose tissue and slow twitch red skeletal muscle. In addition, LPL has been demonstrated in the lung, aorta and kidney medulla [9].

LPL is synthetized within the cells and thereafter transported to the cell surface to finally be located on the surface of the capillary endothelium. The LPL activity is influenced by a range of hormones and differences do exist between the various locations with regard to hormonal regulation.

Catecholamines, glucagon, thyroid hormones (T_3) and glucocorticoids have been found to stimulate LPL in the heart. In skeletal muscle it has been demonstrated that at least adrenaline stimulates LPL. In adipose tissue insulin has been found to activate LPL while adrenaline and ACTH have been found to inhibit the enzyme [85]. In adipose tissue these effects are probably confined to the intracellular control of enzyme synthesis, while once located extracellularly the enzyme activity is no longer under hormonal influence [29]. However, the metabolic status of the organism as such does influence the LPL activity. In the fed state LPL is active in adipose tissue while the enzyme is relatively inactive in skeletal muscle. In the fasted state the enzyme activity pattern is reversed the enzyme being most active in skeletal muscle. Thus, the enzyme activity is regulated depending upon the need for lipid storage or lipid oxidation [8].

Intramuscular Triacylglycerols

A certain amount of the LPL synthetized in heart and skeletal muscle is not transported to the surface of the capillary endothelium but remains located intracellularly to control degradation of intramuscularly located triacylglycerols.

Adrenaline and glucagon have been found to inhibit intracellular LPL in low concentrations while high concentrations led to enzyme activation [36, 83]. The activation of intracellular LPL is much slower than for other enzymes regulated via c-AMP and this has led to the hypothesis that an intracellular inhibitor of LPL exists participating in a multistep regulation of LPL [85].

Lipid Mobilization during Exercise

Methodological Considerations

Classically, lipid oxidation is estimated in vivo from the rates of oxygen uptake and carbon dioxide production, using the stoichiometric equations for total aerobic combustion of glucose and triacylglycerols corrected for protein metabolism determined by urinary excretion of nitrogen. However, it is often assumed that protein metabolism may be regarded as negligible at rest as well as during exercise. During rest, this assumption will introduce an error of 1–1.5% in the calculations of carbohydrate and lipid metabolism, while during exercise the calculations will be practically correct. Other conditions that must be met are respiratory steady state and the absence of other metabolic net processes that can influence oxygen uptake and carbon dioxide production such as the conversion of glucose into fatty acids, a process which may take place in man during excess carbohydrate intake [11, 105].

Since a major fraction of the FA liberated by lipolysis may be re-esterified within the adipocyte, it is important to distinguish between lipolysis and FA mobilization, by which is meant liberation of FA from adipose tissue to the blood.

The rate of lipolysis in adipose tissue can be estimated from the glycerol production since glycerol formed by lipolysis cannot be reutilized in this tissue due to low concentrations of α-glycerokinase [67]. A source of error in estimating the total lipolysis by the glycerol production may be the occurrence of incomplete lipolysis which is not reflected by glycerol production. However, the hormone-sensitive lipase is normally the rate-limiting enzyme in the lipolytic process [3].

An in vivo technique was developed by which a-v differences for glycerol across the inguinal fat pad in the dog could be measured during exercise after cannulation of the external pudendal vein. Concomitantly the adipose tissue blood flow was measured in the pad by ^{133}Xe washout [18]. A disadvantage of this method is that some of the blood in the external pudendal vein is derived from other tissue types, mainly skin. However, it was estimated that this only influenced the estimates of lipolysis to a minor extent [18]. Another disadvantage is that other fat depots may behave differently.

In vivo glycerol and FFA metabolism have been determined by constant isotope dilution techniques. Rate of appearance (R_a) and rate of disappearance (R_d) for glycerol and FFA have been calculated [10, 56, 104]. However, when FFA and glycerol release rates from adipose tissue are calculated from the turnover rates, such calculations generally give higher release rates than can be obtained from a-v difference studies [18, 20, 104] while they are similar to those which can be obtained in vitro [54]. Thus, R_a calculations should be interpreted with care.

FFA and glycerol concentrations in plasma are often taken as indicators of FFA and glycerol turnover. A positive correlation between the plasma concentrations and turnover rates has been demonstrated; however, this relationship may vary with the experimental conditions and the intensity of exercise [2, 10, 53, 90, 104].

Lipid Mobilization from Adipose Tissue during Exercise

Oxidation of FFA may cover up to 90% of the total energy demand late in prolonged exercise (table I). Although never measured directly it can be calculated that the lipolytic rate in adipose tissue obtained during exercise, about 3–6 μmol/(100 g·min) is a considerable fraction of the maximal lipolytic rate obtained under in vitro conditions either in isolated fat cells or in perfused isolated fat pads [20, 54]. In dogs the rate of appearance of glycerol has been found to increase about 4- to 4.5-fold during 3 h of exercise [104]. This includes the glycerol mobilization from adipose tissue, intramuscular lipid stores, and circulating triacylglycerols. A similar 4-fold increase in glycerol mobilization from the subcutaneous adipose tissue in dogs was found during 2 h of exercise by determination of the a-v glycerol difference and flow [18]. The lipolytic rate measured in these pads corresponds to 40–50% of the maximal rate obtained in human subcutaneous tissue in vitro [32].

After the onset of exercise lipolysis intensity increases fast while further increase during the exercise bout is modest [18, 104]. This indicates that lipolysis in adipose tissue is almost constantly stimulated during exercise (fig. 2).

Enhanced sympathoadrenal activity and a depressed circulating insulin concentration are the main stimuli of lipolysis in man during exercise [46]. When exercising dogs are infused with propranolol a strong inhibition of glycerol mobilization immediately takes place and the mobilization is virtually reduced to the pre-infusion rest level [62, 116]. In man, acute β-adrenergic blockade during exercise results in reduced increments in glycerol and FFA plasma concentrations. The effect of exercise on these concentrations was about 1/3 of the effect without blockade [16, 77]. Mc-Leod et al. [77] found decreasing FFA concentrations during β-blockade, and increased concentrations during α-blockade.

Endurance is impaired by β-adrenergic blockade in man [44]. In dog, the impaired working capacity could be restored by infusion of intralipid and heparin, a treatment giving rise to the normal exercise-induced increase in FFA concentration during blockade [78].

The second hormonal mechanism of importance for the increased lipolytic rate during exercise is the depressed insulin concentration, which

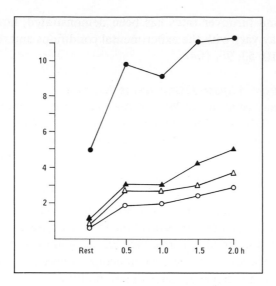

Fig. 2. Simultaneous changes in adipose tissue blood flow (•; ml/(100 g·min)), glycerol (○; μmol/(100 g·min)) and FFA mobilizations (△; μmol/(100 g·min)), and FFA reesterification (▲; μmol/(100 g·min)) during prolonged exercise in dogs [modified after Bülow, 18].

is directly related to work intensity [46]. The concentrations of FFA and glycerol during exercise are higher in fasted and in fat-fed experimental subjects than in normal control subjects, while the insulin concentration is lower [45].

Thyroid function may not be without influence on the lipolytic response during exercise in man. It has been found that the hypothyroid state is associated with increased α-adrenergic responsiveness in human adipose tissue [99]. Similarly, enhanced lipolysis has been demonstrated in the hyperthyroid state by increased β-adrenergic responsiveness [86]. Recently, it has been described that the changes in adrenergic responsiveness is located beyond the adrenoceptor level in hyper- and hypothyroidism [112]. However, there is no evidence that thyroid hormones may be acute regulators of lipolysis and, furthermore, the concentrations of thyroxin and triiodothyronine do not change during exercise [46]. Similarly, there is no experimental evidence for the lipolytic effect in man of other hormones such as glucagon, corticotropin, growth hormone or other peptide hormones in vivo, although an effect can be demonstrated in vitro [54].

The amount of FFA mobilized from the adipocytes depends on the rate of lipolysis and on the competition between the FFA-carrying capacity of the perfusing blood on the one hand, and the process of re-esterification in the adipocytes on the other.

The FFA-carrying capacity is determined by the albumin concentration of the blood, by the blood flow, and by the number of FFA-binding sites on albumin already occupied (i.e. the arterial FFA/albumin molar ratio). Since each albumin molecule can bind a finite number of FA molecules and does so with decreasing affinity [4], an increase in the FFA/albumin ratio must be accompanied by an increase in the concentration of unassociated or 'free' FFA in the blood and tissue water phase. This increase will favor re-esterification at the expense of mobilization by the blood [17, 72, 102]. When isolated adipocytes are exposed to increasing FFA/albumin ratios in the surrounding medium, their FA release is inhibited by an inhibition of lipolysis probably as a consequence of inhibition of adenyl cyclase [13, 42]. Similar inhibition of lipolysis can neither be demonstrated in the isolated, perfused, lipolytically stimulated fat pad [72], nor by a-v difference studies of the inguinal fat pad in situ during exercise [18]. In both these situations, the increase in the arterial FFA/albumin ratio resulted in an enhanced re-esterification. About 2/3 of the FA liberated by lipolysis during exercise is re-esterified in this futile cycle (fig. 2).

During prolonged exercise, adipose tissue blood flow (ATBF) increases, thus facilitating the removal of FA from the tissue. This has been demonstrated in man, dog and rat [20, 69]. In man and dog, ATBF increases about 3-fold in different adipose tissues during prolonged exercise [14–16, 19] (table II). The increase in blood flow has been demonstrated to enhance the FFA mobilization during exercise [19]. It was proposed that the increase in blood flow was secondary to metabolic events connected to lipolysis [16, 20] and recently it has been found that the mediator of this coupling between tissue metabolism and blood flow is adenosine [76], while local denervation does not influence the exercise-induced vasodilatation [22].

High FFA/albumin ratios in the arterial blood have, on the other hand, been shown to increase the vascular resistance in adipose tissue both in perfusion experiments and in the intact dog [21]. The effect is apparent at FFA/albumin ratios of about 3 and increases markedly with this ratio. It has been suggested that this mechanism will inhibit the removal of FFA from adipose tissue, when the arterial FFA/albumin ratio approaches toxic levels [72].

Table II. Adipose tissue blood flow (ml/(100 g·min)) in man and dog during prolonged exercise of moderate intensity [from Bülow et al., 15, 18, 20]

	Rest	1 h	2 h	3 h	4 h
Man					
Subcutis	3.0	4.7	6.2	7.9	7.6
Perirenal	2.3	4.3	8.1	12.8	15.4
Dog					
Subcutis	6.2	9.2	10.3		
Perirenal	10.2	16.4	32.6		
Omental	7.6	11.8	17.5		
Mesenteric	9.5	14.9	30.2		
Pericardial	10.8	16.3	25.3		

Besides the arterial FFA/albumin ratio and the blood flow, the arterial lactate concentration is of importance for the FFA mobilization. In the isolated, perfused adipose tissue from dogs, lactate enhances FA re-esterification without affecting the glycerol release [41]. In exercising dogs, increased lactate concentration decreases the R_{agly} [61].

The importance of precise adjustment of FA mobilization from adipose tissue to the concomitant utilization in working muscle has been emphasized [80]. This adjustment can hardly be achieved by simple neurohormonal regulation of the rate of lipolysis implying that the lipolysis-re-esterification cycle may be instrumental in this adjustment, feedback from the plasma FFA concentration influencing mainly re-esterification while the rate of lipolysis remains uninfluenced [18].

Thus, it was recently proposed [23] that the adjustment of FA mobilization from adipose tissue to the FA utilization in working muscle is accomplished by the following sequence of events (fig. 3). During exercise, lipolysis increases fast to a constant level due to stimulation by sympathoadrenal factors and falling insulin concentrations. Secondary to this, adipose tissue blood flow increases promoting the removal of FAs from the adipocytes. The rate of lipolysis is stimulated far in excess of the need of the working tissues, since about 2/3 of the liberated FAs are re-esterified in the tissue. The amount of FA mobilized from adipose tissue is also in excess of the simultaneous utilization, as evidenced by an increasing arterial FFA concentration. As the FFA/albumin ratio increases, two feedback mechanisms oppose the excessive mobilization: (1) the increased re-

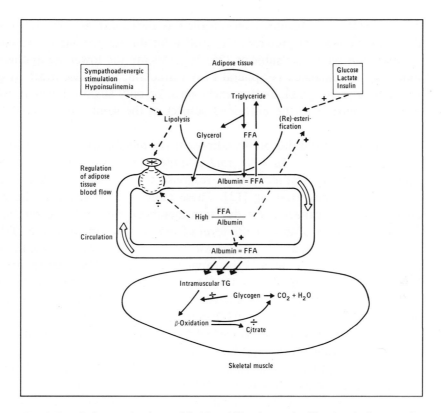

Fig. 3. Regulating mechanisms of lipid mobilization and utilization during exercise in adipose tissue and skeletal muscle.

esterification, and (2) a rising vascular resistance in adipose tissue, opposing the increase in adipose tissue blood flow. As a result, FA mobilization is adapted to utilization, but at an increased FFA concentration level which promotes the FFA uptake in the working muscles. This adjustment is reflected in relatively constant FFA concentrations during the later part of prolonged, constant work.

Effect of Training on Adipocyte Function
The effect of endurance training on lipid metabolism during exercise has provided varying information. It is well established that the plasma catecholamine response to moderate exercise is lower after endurance

training [55, 115]. Insulin concentration is decreased, too, by training [7]. Trained subjects preferentially utilize fat during prolonged exercise compared to sedentary subjects [26, 59]. Despite the lower sympathetic tone and the enhanced peripheral FFA extraction, the concentrations of FFA and glycerol in blood are unchanged, or only slightly lower in the trained compared to the untrained subject at the same work level [52, 115].

The effect of training on the adipose tissue metabolism has been studied mainly in the rat. In adult rats of either sex, it was found that the lipolytic capacity of adipocytes in response to adrenaline markedly increased by exercise training [12]. These authors proposed that the increased lipolytic capacity was due to a metabolic step distal to the hormonal receptor, probably at the level of the protein kinase or the HSL. However, in Oscai et al. [84] a decrease was found in protein kinase activity after exercise training, and the capacity to activate HSL was reduced. Thus, it was concluded that the lipolytic capacity in the untrained state is sufficiently great to cover the increased oxidation of FFA after training without an adaptive increase in HSL activity. Similarly, it was not possible to demonstrate any effect on the cAMP system in adipose tissue from male rats after endurance training [87].

This is in contrast to the findings of Williams and Bishop [114] of an enhanced adenylate cyclase activity after epinephrine stimulation of adipocytes from endurance-trained rats. The number of β-adrenergic receptors was not changed.

From the evidence cited above, it seems that physical training is associated with an increased sensitivity to β-adrenergic agonists without alteration in the number of receptors.

The effect of insulin following training has been studied in isolated adipocytes from rats. Strenuous training resulted in an increased transport of 3-0-methylglucose into adipocytes under the influence of insulin [111]. The maximally insulin-stimulated incorporation of glucose carbon into lipids was also increased by training. Similar results are reported by Craig et al. [28] who found that 2-deoxyglucose uptake and glucose oxidation were higher in fat cells from trained compared with sedentary rats. At maximal stimulation, glucose uptake and oxidation was increased about 6-fold in the trained rats. The effect could not be explained by an increased insulin binding, although the adipocytes from exercise-trained rats bound more insulin than those from sedentary controls. However, even these sedentary rats can bind about 6 times more insulin than required for the maximal response. This is in keeping with the concept of 'spare' receptors [47]. Wardzala et al. [113] found that the increase in glucose metabolism could be demonstrated in CO_2 production as well as in de novo synthesis of fatty acids and in glycerol production for triacylglycerol synthesis, confirming the findings by York et al. [117] that insulin may influence adipose cell metabolism beyond the glucose transport.

The combined effects of increased lipolytic sensitivity to β-adrenergic stimulation and increased insulin sensitivity giving increased glucose uptake and increased lipogenesis suggest that the main effect of physical training in rat adipose tissue is to increase the lipid mobilizing capacity (β-adrenergic effect) and to enhance the ability to replenish the emptied triacylglycerol store (insulin effect).

In man, training has also been suggested to be associated with increased β-adrenergic sensitivity in adipose tissue [68]. This increase in lipolytic activity in response to adrenaline infusion following endurance training was confirmed by Martin and Bockman [76]. On the other hand, it was found that only 4 days after cessation of training, this enhanced lipolytic effect was lost. As in the rat, exercise training augments adrenaline-stimulated lipolysis in isolated suprailiac subcutaneous adipocytes from man [32]. The adaptive increase in lipolysis seems to be maximal within about 4 months after the onset of training [33]. There are no available data on adipose tissue perfusion before and after training; neither are any data available on the dependency of the FFA re-esterification rate on the arterial FFA-binding capacity after training.

Lipolysis of Circulating Lipoproteins during Exercise
The plasma triacylglycerols (TG) are found in the chylomicrons originating from intestinal absorption of lipids and in the very low density lipoproteins (VLDL) originating from the liver. The role of circulating TG as energy source during exercise has been considered as relatively small. In man, a moderate work load did not alter the plasma concentration of TG while prolonged heavy exercise decreased the level [24, 109].

However, Terjung et al. [108] have recently suggested that circulating TG may represent a potential source of FA for β-oxidation in exercising muscle, at least in the dog. By the isotope dilution technique, these authors found a high absolute TG turnover rate in the postprandial resting state, approximately 3 times that of the circulating FFA. During exercise there was a tendency to increase in the fractional turnover of TG. However, since TG concentration is unchanged or even decreased in this situation this may represent an unchanged absolute turnover. Thus, in contrast to the greater turnover of FFA during exercise mainly due to increased plasma concentrations, changes in TG turnover are mainly influenced or even limited by factors influencing tissue uptake. The greater uptake of FA derived from TG in exercising dogs could be due to two factors. Primarily, it could be due to enhanced LPL activity during exercise in the muscle tissue with direct uptake of the released FFA as a result. Secondarily, it could be due to uptake of TG-derived FFA circulating in the FFA pool thus being taken up parallel to FFA derived from other sources. The importance of these two possibilities has not been evaluated. In favor of the first may be the finding that uptake of TG-derived FFA is enhanced before the turnover of plasma FFA is increased during short-term exercise [108]. In favor of a significance for the second mechanism may be the finding by Bergman et al. [6] that only 50–60% of FFA released from TG in perfused dog hindquarters were taken up directly into the tissue.

Training has been found to lower circulating TG concentrations in man [70]. However, since turnover was not measured this effect could be due to decreased output of VLDL from the liver as well as increased uptake in adipose tissue, heart and skeletal muscle.

Lipolysis of Intramuscular Triacylglycerols during Exercise

It has been calculated that about half of the total FFA being oxidized during prolonged moderate exercise is derived from circulating FFA while the other half is derived from the endogenous TG in the muscle [37, 43, 88]. It is well established that the concentration of endogenous TG concentration decreases in skeletal muscle [37, 106]. The concentration of intramuscular TG is normally around 5–15 mmol/kg wet weight but with great variation [37]. This concentration decreases between 25 and 50% during an exercise bout depending on the mode of exercise [37]. This corresponds to an average mobilization of about 2–5 µmol/(100 g·min). This is a lipolytic intensity similar to that of adipose tissue but fractionally it is considerably higher since the TG concentration in adipose tissue is about 400–800 mmol/kg wet weight.

On the other hand, it has been found that intracellular TG content is normalized 1–3 h after an exercise bout [51]. LPL activity does increase during prolonged exercise, i.e. exercise of more than 1 hour's duration. In skeletal muscle, the LPL activity rose about 2-fold and in adipose tissue the enzyme activity rose by about 20% [107].

Training does influence LPL activity. Adipose tissue LPL activity was on average 70% higher in trained subjects than in untrained controls [74].

Lipid Utilization during Exercise

The utilization of FFA during exercise depends on work intensity, state of physical training, and diet.

Uptake of FFA from blood to cells is a simple concentration-dependent process not requiring energy supply. The transfer from the carrier proteins in blood to the cell is only dependent on the binding affinity for FFA of the plasma proteins versus the affinity of binding sites in the bilayer plasma membrane. The transfer from blood to cell is rapid. On the other hand, it seems that this physical-chemical partitioning of FFA between plasma protein and the plasma membranes provides a sufficient regulatory capacity to prevent uptake that exceeds the intracellular metabolic capacity [82]. It has similarly been demonstrated by turnover studies that the FFA turnover in plasma is dependent on the FFA concentration. As mentioned above under the regulation of FFA mobilization from adipose tissue, this leaves this mobilization as the major determinant of muscle utilization of FFA. Once located on the inside of the plasma membrane the FFA are converted to fatty acyl-CoA, and as such they may either be reesterified and stored as triacylglycerol or transported over the mitochondrial membrane into the mitochondrial matrix where they are β-oxidized. The transport of the fatty acyl-CoA complex over the mitochondrial membrane is entirely dependent on the presence of carnitine [40]. Carnitine may be derived from two sources, synthesis and dietary intake. About 75% if not all of the carnitine needed in the body is derived from synthesis in the liver. The uptake of carnitine into muscle is accomplished by active transport. Thereafter, carnitine is complexed to different forms of acylcarnitine transferases with maximal affinity for short-chain (carnitine acetyl transferase), medium-chain (carnitine octanoyl transferase), and long-chain (carnitine palmityl transferase) FA. These acyl-carnitine transferases are located on the mitochondrial membrane in one form on the outer membrane which converts fatty acyl-CoA to acyl-carnitine, and in another form located on the inner membrane converting acyl-carnitine back to fatty acyl-CoA which can then be β-oxidized (fig. 4). If carnitine is not available, disorders of lipid metabolism and myopathies will follow [34].

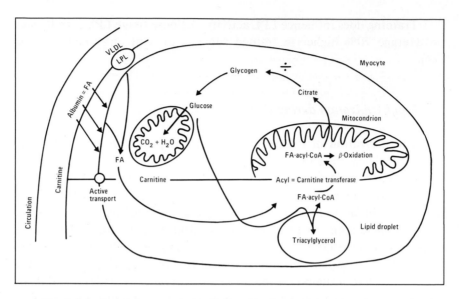

Fig. 4. Principal factors regulating lipid utilization in skeletal muscle.

FAs are predominantly oxidized in slow-twitch red fibers which are mainly activated during exercise of moderate intensity [101]. Thus, the contribution of lipid oxidation to total oxidative metabolism is dependent on the relative work load. During heavy work carbohydrates contribute with about 80% of the energy, while during prolonged exercise of moderate intensity lipid combustion may cover up to 90% of the substrates metabolized [90] (table I).

The regulation of fuel selection during exercise is not fully understood. In 1963, Randle et al. [91] proposed the FFA-glucose cycle as the key regulating mechanism. The biochemical basis of this concept is that oxidation of FFA and ketone bodies gives rise to increased acetyl-CoA concentration and increased acetyl-CoA/CoA ratio which inhibits pyruvate dehydrogenase. High concentrations of citrate derived from acetyl-CoA will inhibit phosphofructokinase leading to accumulation of glucose-6-phosphate and subsequently inhibition of hexokinase and thus glucose oxidation. Glucose on the other hand, will influence the FFA metabolism via the lipolysis/reesterification process. The operation of the glucose-FFA cycle has been confirmed in the slow-twitch red fibers [30, 73] in isolated soleus muscles from mice and rats. Similarly, it has been found that oleate inhibits glucose uptake in the perfused rat hindquarter [94]. Newsholme et al. [79] found similar effects. However, others have failed to demonstrate an inhibitory effect of FFA on glucose metabolism [49, 93].

It has been suggested that the elevated FFA concentration during prolonged exercise of moderate intensity might have a sparing effect on the muscular glycogen utilization [25, 97]. However, in recent studies in man during prolonged exercise of moderate intensity it has not been possible to demonstrate any effect of elevated FFA concentration, medium chain as well as long chain, on oxidative fuel selection [31, 63, 92] raising doubt of a physiological role of the glucose-FA cycle in this situation in man. On the other hand, when the lipid mobilization is inhibited by administration of nicotinic acid, endurance is reduced [16].

The effect of diet on fuel selection during exercise has been found to be accomplished through the size of the muscular glycogen stores. When the glycogen stores are full the oxidation of FFA in muscle is inhibited compared to a situation in which these stores are emptied [48]. The increase in circulating FFA concentration during exercise is smaller after a high-carbohydrate diet than after a high-fat diet [45], and glucose infusion during exercise likewise gives rise to a smaller increase in FFA and glycerol concentrations compared to control conditions [16]. Similarly, ingestion of glucose during prolonged exercise diminishes lipolysis [1].

Not only the availability of FAs versus glucose determines the selection of fuels during exercise, the environment is also of importance. Exercise in the cold enhances lipid oxidation compared to exercise in neutral temperatures [60, 110].

Effect of Endurance Training on Lipid Utilization

As stated above, the oxidation of FAs is determined by 2 main factors at a given metabolic rate. These are the availability of FFA, determined by FFA mobilization and capacity of the tissues to oxidize FFA.

Since FFA, even in the untrained state, are mobilized beyond the need to cover oxidation evidenced by the increasing FFA concentration, and since the FFA concentration at a given relative work load is lower in the trained state compared to the untrained state [64, 95], this cannot explain the greater capacity to oxidize FFA after endurance training [58, 100]. Thus, the enhanced lipid combustion must be due to changes in the capacity to oxidize FFA mainly in skeletal muscle. It has been demonstrated that enzymatic adaptation takes place after endurance training. However, the in vitro findings do not correlate well to the in vivo findings [27].

Metabolism of Ketone Bodies during Exercise

Ketone bodies (acetoacetate and β-hydroxybutyrate) are formed in the liver during degradation of FFA and they are then given off into the circulating blood. In normal man the concentration of ketone bodies in blood is rather low, in the range of 0–0.2 mmol/l, while during fasting or deranged diabetes mellitus it may increase to about 4–6 mmol/l. The rate of ketogenesis is linearly correlated to the FFA load on the liver [103].

During short bouts of exercise in healthy subjects the ketone bodies contribute very little to the metabolism of the exercising muscles and probably not more than during rest [53, 71, 96]. Recently, it has been found by isotope dilution studies that the response of ketone body metabolism is dependent on the level of ketonemia during prolonged exercise of moderate intensity [38, 39]. During low levels of ketonemia, i.e. <0.6 mmol/l, the rate of appearance (R_a) increases correlated to the increase in FFA concentration and thus as a result of increased lipolysis. During high levels of ketonemia, i.e. >3 mmol/l, this stimulatory effect of exercise disappeared and R_a was unchanged or even decreased. Rate of disappearance (R_d) similarly increased for low levels of ketonemia while it remained unchanged or decreased during high levels. The metabolic clearance rate showed similar changes. However, even at high concentrations of ketone bodies the oxidation in skeletal muscle cannot account for more than maximally 7% of the skeletal muscle metabolism.

Both the diminishing or abolishing effect of exercise on liver ketone body production and on the metabolic clearance rate (i.e. muscular extraction) may serve to keep the ketone body concentration at a constant high level and save them for utilization in nonmuscular tissues, principally the brain, in these metabolic states in which glucose availability is low.

References

1 Ahlborg, G.; Felig, P.: Influence of glucose ingestion on fuel-hormone response during prolonged exercise. J. appl. Physiol. *41:* 683–688 (1976).
2 Ahlborg, A.; Felig, P.; Hagenfeldt, L.; Hendler, R.; Wahren, J.: Substrate turnover during prolonged exercise in man. Splanchnic and leg metabolism of glucose, free fatty acids, and amino acids. J. clin. Invest. *53:* 1080–1090 (1974).
3 Arner, P.; Östman, J.: Mono- and diacylglycerols in human adipose tissue. Biochim. biophys. Acta *369:* 209–221 (1974).
4 Ashbrook, J.D.; Spector, A.A.; Santos, E.C.; Fletcher, J.E.: Long chain fatty acid binding to human plasma albumin. J. biol. Chem. *6:* 2333–2338 (1975).

5 Belfrage, P.: Hormonal control of lipid degradation; in Cryer, Van, New perspectives in adipose tissue. Structure function and development, pp. 121–144 (Butterworths, London 1985).

6 Bergman, E.N.; Havel, R.J.; Wolfe, B.M.; Bøhmer, T.: Quantitative studies of the metabolism of chylomicron triglycerides and cholesterol by liver and extrahepatic tissues of sheep and dogs. J. clin. Invest. 53: 1831–1839 (1971).

7 Björntorp, P.; Jounge, K. de; Sjöström, L.; Sullivan, L.: The effect of physical training on insulin production in obesity. Metabolism 19: 631–638 (1970).

8 Borensztajn, J.; Robinson, D.S.: The effect of fasting on the utilization of chylomicron triglyceride fatty acids in relation to clearing factor lipase (lipoprotein lipase) released by heparin in the perfused rat heart. J. Lipid Res. 11: 111–117 (1970).

9 Borensztajn, J.: Lipoprotein lipase; in Scann, Wissler, Getz, The biochemistry of atherosclerosis, pp. 231–245 (Marcel Dekker, New York 1979).

10 Bortz, W.M.; Paul, P.; Haff, A.C.; Holmes, W.L.: Glycerol turnover and oxidation in man. J. clin. Invest. 51: 1537–1546 (1972).

11 Bray, G.A.: Lipogenesis in human adipose tissue. Some effects of nibbling and gorging. J. clin. Invest. 51: 537–548 (1972).

12 Bukowiecki, L.; Lupien, J.; Follea, N.; Paradis, A.; Richard, D.; LeBlanc, J.: Mechanism of enhanced lipolysis in adipose tissue of exercise-trained rats. Am. J. Physiol. 239: E422–E429 (1980).

13 Burns, G.W.; Langley, P.E.; Robinson, G.A.: Site of free-fatty-acid inhibition of lipolysis by human adipocytes. Metabolism 24: 265–276 (1975).

14 Bülow, J.; Madsen, J.: Adipose tissue blood flow during prolonged, heavy exercise. Pflügers Arch. 363: 231–234 (1976).

15 Bülow, J.; Madsen, J.: Human adipose tissue blood flow during prolonged exercise. Part II. Pflügers Arch. 376: 41–45 (1978).

16 Bülow, J.: Human adipose tissue blood flow during prolonged exercise. III. Effect of β-adrenergic blockade, nicotinic acid and glucose infusion. Scand. J. clin. Lab. Invest. 41: 415–424 (1981).

17 Bülow, J.; Madsen, J.: Influence of blood flow on fatty acid mobilization from lipolytically active adipose tissue. Pflügers Arch. 390: 169–174 (1981).

18 Bülow, J.: Subcutaneous adipose tissue blood flow and triacylglycerol mobilization during prolonged exercise in dogs. Pflügers Arch. 392: 230–234 (1982).

19 Bülow, J.; Tøndevold, E.: Blood flow in different adipose tissue depots during prolonged exercise in dogs. Pflügers Arch. 392: 235–238 (1982).

20 Bülow, J.: Adipose tissue blood flow during exercise. Dan. med. Bull. 30: 85–100 (1983).

21 Bülow, J.; Madsen, J.; Astrup, A.; Christensen, N.J.: The effect of high free fatty acid/albumin molar ratios on the perfusion of adipose tissue in vivo. Acta physiol. scand. 125: 661–667 (1985).

22 Bülow, J.; Madsen, J.: Dog subcutaneous adipose tissue blood flow during exercise after surgical denervation. Acta physiol. scand. 128: 471–474 (1986a).

23 Bülow, J.; Madsen, J.: Regulation of lipid mobilization during exercise. Scand. J. Sports Sci. 8: 19–26 (1986b).

24 Carlson, L.A.; Mossfeldt, F.: Acute effects of prolonged heavy exercise on the concentration of plasma lipids and lipoproteins in man. Acta physiol. scand. 62: 51–59 (1964).

25 Costill, D.L.; Coyle, E.; Dalsky, G.; Evans, W.; Fink, W.; Hoopes, D.: Effects of elevated plasma FFA and insulin on muscle glycogen usage during exercise. J. appl. Physiol. *43:* 695–699 (1977).

26 Costill, D.L.; Fink, W.J.; Getchell, L.H.; Ivy, J.L.; Witzmann, F.A.: Lipid metabolism in skeletal muscle of endurance-trained males and females. J. appl. Physiol. *47:* 787–791 (1979).

27 Costill, D.L.; Sherman, W.M.; Essig, D.A.: Metabolic responses and adaptations to endurance running; in Poortmans, Niset, Biochemistry of exercise IV-A, pp. 33–45 (University Park Press, Baltimore 1981).

28 Craig, B.W.; Hammans, G.T.; Garthwaite, S.M.; Garett, L.; Holloszy, J.O.: Adaptation of fat cells to exercise: response of glucose uptake and oxidation to insulin. J. appl. Physiol. *51:* 1500–1506 (1981).

29 Cryer, A.: Lipoproteins and adipose tissue; in Cryer, Van, New perspectives in adipose tissue, structure, function and development, pp. 183–198 (Butterworths, London 1985).

30 Cuendet, G.S.; Loten, E.G.; Renold, A.E.: Evidence that the glucose fatty acid cycle is operative in isolated skeletal (soleus) muscle. Diabetologia *12:* 336 (1975).

31 Decombaz, J.; Arnaud, M.H.; Milou, H.; Moesch, H.; Philippossian, G.; Tehlin, A.L.; Howald, H.: Energy metabolism of medium-chain triglycerides versus carbohydrate during exercise. Eur. J. appl. Physiol. occup. Physiol. *52:* 9–14 (1983).

32 Després, J.P.; Bouchard, C.; Savard, R.; Tremblay, A.; Marcotte, M.; Theriault, G.: Level of physical fitness and adipocyte lipolysis in humans. J. appl. Physiol. *56:* 1157–1161 (1984a).

33 Després, J.P.; Bouchard, C.; Savard, R.; Tremblay, A.; Marcotte, M.; Theriault, G.: The effect of a 20-week endurance training program on adipose-tissue morphology and lipolysis in men and women. Metabolism *33:* 235–239 (1984b).

34 Di Mauro, S.; Trevisan, C.; Hays, A.: Disorders of lipid metabolism in muscle. Muscle Nerve *3:* 369–388 (1980).

35 Dole, V.P.: A relation between non-esterified fatty acids in plasma and the metabolism of glucose. J. clin. Invest. *35:* 150–154 (1956).

36 Eaton, R.P.: Glucagon and lipoprotein regulation in man; in Foá, Bajaj, Foá, Glucagon: its role in physiology and clinical medicine, pp. 533–550 (Springer, New York 1977).

37 Essén, B.: Intramuscular substrate utilization during prolonged exercise. N.J. Acad. Sci. *301:* 30–44 (1977).

38 Féry, F.; Balasse, O.: Ketone body turnover during and after exercise during in overnight-fasted and starved humans. Am. J. Physiol. *245:* E318–E325 (1983).

39 Féry, F.; Balasse, O.: Response of ketone body metabolism to exercise during transition from postabsorptive to fasted state. Am. J. Physiol. *250:* E495–E501 (1986).

40 Fritz, J.B.: Carnitine and its role in fatty acid metabolism. Adv. Lipid Res. *1:* 285–334 (1963).

41 Fredholm, B.B.: The effect of lactate in canine subcutaneous adipose tissue in situ. Acta physiol. scand. *81:* 110–123 (1971).

42 Fredholm, B.B.: Local regulation of lipolysis in adipose tissue by fatty acids, prostaglandines and adenosine. Med. Biol. *56:* 249–261 (1978).

43   Fröberg, S.O.; Mossfeldt, F.: Effect of prolonged strenuous exercise on the concentration of triglycerides, phospholipids and glycogen in muscle of man. Acta physiol. scand. *82:* 167–171 (1971).

44   Galbo, H.; Holst, J.J.; Christensen, N.J.; Hilsted, J.: Glucagon and plasma catecholamines during beta-receptor blockade in exercising man. J. appl. Physiol. *40:* 855–863 (1976).

45   Galbo, H.; Holst, J.J.; Christensen, N.J.: The effect of different diets and of insulin on the hormonal response to prolonged exercise. Acta physiol. scand. *107:* 19–32 (1979).

46   Galbo, H.: Hormonal and metabolic adaptation to exercise (Thieme, Stuttgart 1983).

47   Gammeltoft, S.; Gliemann, J.: Binding and degradation of 125I-labeled insulin by isolated rat fat cells. Biochim. biophys. Acta *320:* 16–32 (1973).

48   Gollnick, P.D.; Pernow, B.; Essén, B.; Jansson, E.; Saltin, B.: Availability of glucogen and plasma FFA for substrate utilization in leg muscle of man during exercise. Clin. Physiol. *1:* 27–42 (1981).

49   Goodman, M.N.; Berger, M.; Ruderman, N.B.: Glucose metabolism in rat skeletal muscle at rest. Effect of starvation, diabetes, ketone bodies and free fatty acids. Diabetes *23:* 881–888 (1974).

50   Gordon, R.S.; Cherkes, A.: Unesterified fatty acid in human blood plasma. J. clin. Invest. *35:* 206–212 (1956).

51   Gorski, J.; Kiryluk, T.: The post-exercise recovery of triglycerides in rat tissues. Eur. J. appl. Physiol. *45:* 33–41 (1980).

52   Gyntelberg, F.; Rennie, M.J.; Hickson, R.C.; Holloszy, J.O.: Effect of training on the response of plasma glucagon to exercise. J. appl. Physiol. *43:* 302–305 (1977).

53   Hagenfeldt, L.: Metabolism of free fatty acids and ketone bodies during exercise in normal and diabetic man. Diabetes *28:* suppl. 1, pp. 66–70 (1979).

54   Hales, C.N.; Luzio, J.P.; Liddle, K.: Hormonal control of adipose-tissue lipolysis. Biochem. Soc. Symp. *43:* 97–135 (1978).

55   Hartley, L.H.; Mason, J.W.; Hogan, R.P.; Jones, L.G.; Kotchen, T.A.; Mongey, E.H.; Wherry, F.E.; Pennington, L.L.; Ricketts, P.T.: Multiple hormonal responses to prolonged exercise in relation to physical training. J. appl. Physiol. *33:* 607–610 (1972).

56   Havel, R.J.; Carlson, L.A.: Comparative turnover rates of free fatty acids and glycerol in blood of dogs under various conditions. Life Sci. *9:* 651–658 (1963).

57   Heindel, J.J.; Orci, L.; Jeanrenaud, B.: Fat mobilization and its regulation by hormones and drugs in white adipose tissue; in Masoro, Pharmacology of lipid transport and atherosclerotic processes, pp. 175–373 (Pergamon Press, Oxford 1975).

58   Hermansen, L.; Hultman, E.; Saltin, B.: Muscle glycogen during prolonged severe exercise. Acta physiol. scand. *71:* 129–139 (1967).

59   Holloszy, J.O.; Booth, F.W.: Biochemical adaptations to endurance exercise in muscle. A. Rev. Physiol. *38:* 273–291 (1976).

60   Hurley, B.F.; Haymes, E.M.: The effects of rest and exercise in the cold on substrate mobilization and utilization. Aviat. Space envir. Med. *53:* 1193–1197 (1982).

61   Issekutz, B.; Miller, H.: Plasma free fatty acids during exercise and the effect of lactic acid. Proc. Soc. exp. Biol. Med. *110:* 237–239 (1962).

62 Issekutz, B.: Role of beta-adrenergic receptors in mobilization of energy sources in exercising dogs. J. appl. Physiol. *44:* 869–876 (1978).

63 Ivy, J.L.; Costill, D.L.; Fink, W.J.; Maglisho, E.: Contribution of medium and long chain triglycerides intake to energy metabolism during prolonged exercise. Int. J. Sports Med. *1:* 15–20 (1980).

64 Johnson, R.H.; Walton, J.L.; Krebs, H.A.; Williamson, D.H.: Metabolic fuels during and after severe exercise in athletes and non-athletes. Lancet *ii:* 452–455 (1969).

65 Kather, H.; Simon, B.: Catecholamine-sensitive adenylate cyclase of human fat cell ghosts. A comparative study using different beta-adrenergic agents. Metabolism *24:* 1179–1182 (1977).

66 Kather, H.; Pries, J.; Schrader, V.; Simon, B.: Inhibition of human fat cell adenylate cyclase mediated via alpha-adrenoceptors. Eur. J. clin. Invest. *10:* 345–348 (1980).

67 Koschinsky, T.; Gries, F.A.: Glycerin-Kinase und Lipolyse des menschlichen Fettgewebes in Abhängigkeit vom relativen Körpergewicht. Hoppe-Seyler's Z. physiol. Chem. *352:* 430 (1971).

68 Krotkiewski, M.; Mandroukas, K.; Morgan, L.; William-Olsson, T.; Feurle, G.E.; von Schenck, H.; Björntorp, P.; Sjöström, L.; Smith, U.: Effects of physical training on adrenergic sensitivity in obesity. J. appl. Physiol. *55:* 1811–1817 (1983).

69 Larsen, T.; Myhre, K.; Vik-Mo, H.; Mjös, O.D.: Adipose tissue perfusion and fatty acid release in exercising rats. Acta physiol. scand. *113:* 111–116 (1981).

70 Lehtonen, A.; Viikari, J.: Serum triglycerides and cholesterol and high-density lipoprotein cholesterol in highly physically active men. Acta med. scand. *204:* 111–114 (1978).

71 Lyngsøe, J.; Clausen, J.P.; Trap-Jensen, J.; Sestoft, L.; Schaffalitzky de Muckadell, O.; Holst, J.J.; Nielsen, S.L.; Rehfeld, J.F.: Exchange of metabolites in the leg of exercising juvenile diabetic subjects. Clin. Sci. mol. Med. *55:* 53–80 (1978).

72 Madsen, J.; Bülow, J.; Nielsen, N.E.: Feed back regulation of fatty acid mobilization by arterial free fatty acid concentration. Acta physiol. scand. *127:* 161–166 (1986).

73 Maizels, E.Z.; Ruderman, N.B.; Goodman, M.N.; Lan, D.: Effect of acetoacetate on glucose metabolism in the soleus and extensor digitorum longus muscles of the rat. Biochem J. *162:* 557–568 (1977).

74 Marniemi, J.; Peltonen, P.; Vuori, T.; Hietanen, E.: Lipoprotein lipase of human postheparin plasma and adipose tissue in relation to physical training. Acta physiol. scand. *110:* 131–135 (1980).

75 Martin, W.H., III; Coyle, E.F.; Joyner, M.; Santeusanio, D.; Ehsani, A.A.; Holloszy, J.O.: Effects of stopping exercise training on epinephrine-induced lipolysis in humans. J. appl. Physiol. *56:* 845–848 (1984).

76 Martin, S.E.; Bockman, E.L.: Adenosine regulates blood flow and glucose uptake in adipose tissue in dogs. Am. J. Physiol. *250:* H1127–H1135 (1986).

77 McLeod, A.A.; Brown, J.E.; Kitchell, B.B.; Sedor, F.A.; Kuhn, C.; Shand, D.G.; Williams, R.S.: Hemodynamic and metabolic responses to exercise after adrenoceptor blockade in humans. J. appl. Physiol. *56:* 716–722 (1984).

78 Nazar, K.; Brzezinska, Z.; Kozlowski, S.: Sympathetic activity during prolonged exercise in dogs. Control of energy substrate utilization; in Poortmans, Howald,

Metabolic adaptation to prolonged physical exercise, pp. 204–210 (Birkhäuser, Basel 1975).

79 Newsholme, E.A.; Sugden, P.H.; Williams, T.: Effect of citrate on the activities of phosphofructo kinase from nervous and muscle tissues from different animals and its relationship to the regulation of glycolysis. Biochem. J. *166:* 123–129 (1977).

80 Newsholme, E.A.: The control of fuel utilization by muscle during exercise and starvation. Diabetes *28:* suppl. 1, pp. 1–7 (1979).

81 Nilsson, N.Ö.; Strålfors, P.; Fredrikson, G.; Belfrage, P.: Regulation of adipose tissue lipolysis: effects of noradrenaline and insuline on phosphorylation of hormone-sensitive lipase and on lipolysis in intact rat adipocyte. FEBS Lett. *111:* 125–130 (1980).

82 Noy, A.; Donnelly, T.M.; Zakim, D.: Physical-chemical model for the entry of water-insoluble compounds into cells. Studies of fatty acid uptake by the liver. Biochemistry *25:* 2013–2021 (1986).

83 Oscai, L.B.: Role of lipoprotein lipase in regulating endogenous triacylglycerols in rat heart. Biochem. biophys. Res. Commun. *91:* 227–232 (1979).

84 Oscai, L.B.; Caniso, R.A.; Wergeles, A.C.; Palmer, W.K.: Exercise and the cAMP system in rat adipose tissue. I. Lipid mobilization. J. appl. Physiol. *50:* 250–254 (1981).

85 Oscai, L.B.; Palmer, W.K.: Cellular control of triacylglycerol metabolism. Exerc. Sports Sci. Rev. *11:* 1–23 (1983).

86 Östman, J.; Arner, P.; Bolinger, J.; Engfeldt, P.; Wennlund, A.: Regulation of lipolysis in hyperthyroidism. Int. J. Obes. *5:* 665–670 (1981).

87 Palmer, W.K.; Kalina, C.A.; Studney, T.A.; Oscai, L.B.: Exercise and the cAMP system in rat adipose tissue. II. Nucleotide catabolism. J. appl. Physiol. *50:* 254–258 (1981).

88 Paul, P.; Issekutz, B.: Role of extramuscular energy sources in the metabolism of the exercising dog. J. appl. Physiol. *22:* 615–622 (1967).

89 Paul, P.: FFA metabolism of normal dogs during steady-state exercise at different work loads. J. appl. Physiol. *28:* 127–132 (1970).

90 Paul, P.: Effects of long lasting physical exercise and training on lipid metabolism; in Howald, Poortmans, Metabolic adaptation to prolonged physical exercise, pp. 156–193 (Birkhäuser, Basel 1975).

91 Randle, P.J.; Garland, P.B.; Hales, C.N.; Newsholme, E.A.: The glucose-fatty acid cycle: its role in insulin sensitivity and the metabolic disturbances of diabetes mellitus. Lancet *i:* 785–789 (1963).

92 Ravussin, E.; Bogardus, C.; Scheidegger, K.; LaGrange, B.; Horton, E.; Horton, E.S.: Effect of elevated FFA on carbohydrate and lipid oxidation during prolonged exercise in humans. J. appl. Physiol. *60:* 863–900 (1986).

93 Reimer, F.; Loffler, G.; Hennig, G.; Wieland, O.H.: The influence of insulin on glucose and fatty acid metabolism in isolated perfused rat hind quarter. Hoppe-Seyler's Z. physiol. Chem. *356:* 1055–1066 (1975).

94 Rennie, M.J.; Holloazy, J.O.: Inhibition of glucose uptake and glycogenolysis by availability of oleate in well oxygenated perfused skeletal muscle. Biochem. J. *168:* 161–170 (1977).

95 Rennie, M.J.; Jurnett, S.; Johnson, R.H.: The metabolic effects of strenuous exercise. A comparison between untrained subjects and racing cyclists. Q. Jl exp. Physiol. *59:* 201–212 (1974).

96 Rennie, M.J.; Park, E.M.; Sulaiman, W.R.: Uptake and release of hormones and metabolites by tissues of exercising leg in man. Am. J. Physiol. *231:* 967–973 (1976).
97 Rennie, M.J.; Winder, W.W.; Holloszy, J.O.: A sparing effect of increased plasma fatty acids on muscle and liver glycogen content in the exercising rat. Biochem. J. *156:* 647–655 (1976).
98 Rosell, S.; Belfrage, E.: Blood circulation in adipose tissue. Physiol. Rev. *59:* 1078–1104 (1979).
99 Rosenquist, U.: Inhibition of noradrenalin induced lipolysis in hypothyroid subjects by increased alpha adrenergic responsiveness. An effect mediated through the reduction of cyclic AMP levels in adipose tissue. Acta med. scand. *192:* 353–359 (1972).
100 Saltin, B.; Karlsson, J.: Muscle glycogen utilization during work of different intensities; in Pernow, Saltin, Muscle metabolism during exercise, pp. 289–299 (Plenum Press, New York 1971).
101 Saltin, B.: Muscle fibre recruitment and metabolism in prolonged exhaustive dynamic exercise. Ciba Found. Symp. *82:* 41–58 (1982).
102 Scow, R.O.: Perfusion of isolated adipose tissue: FFA release and blood flow in rat parametrial fat body; in Handbook of physiology, sect. 5: Adipose tissue, pp. 437–453 (American Physiological Society, Washington 1965).
103 Sestoft, L.; Trap-Jensen, J.; Lyngsøe, J.; Clausen, J.P.; Holst, J.J.; Nielsen, S.L.; Rehfeld, J.F.; Schaffalitzky de Muckadell, O.: Regulation of gluconeogenesis and ketogenesis during rest and exercise in diabetic subjects and normal man. Clin. Sci. mol. Med. *53:* 411–418 (1977).
104 Shaw, W.A.S.; Issekutz, T.B.; Issekutz, B.: Interrelationship of FFA and glycerol turnovers in resting and exercising dogs. J. appl. Physiol. *39:* 30–36 (1975).
105 Sjöström, L.; Björntorp, P.; Mansson, J.E.: An optimal assay system for subcellular determination of denovo fatty acid synthesis in human adipose tissue. Scand. J. clin. Lab. Invest. *31:* 191–204 (1973).
106 Spriet, L.L.; Heigenhauser, G.J.F.; Jones, N.L.: Endogenous triacylglycerol utilization by rat skeletal muscle during tetanic stimulation. J. appl. Physiol. *60:* 410–415 (1986).
107 Taskinen, M.-R.; Nikkilä, E.A.; Rekunen, S.; Gordin, A.: Effect of acute vigorous exercise on lipoprotein lipase activity of adipose tissue and skeletal muscle in physically active men. Artery *6:* 471–483 (1980).
108 Terjung, R.L.; Budohoski, L.; Nazar, K.; Kobrún, A.; Kaciuba-Uscilko, H.: Chylomicron triglyceride metabolism in resting and exercising fed dogs. J. appl. Physiol. *52:* 815–820 (1982).
109 Thompson, P.D.; Cullinane, E.; Henderson, L.O.; Herbert, P.A.: Acute effects of prolonged exercise on serum lipids. Metabolism *29:* 662–665 (1980).
110 Timmons, B.A.; Aranjo, J.; Thomas, T.R.: Fat utilization enhanced by exercise in a cold environment. Med. Sci. Sports Exerc. *17:* 673–678 (1985).
111 Vinten, J.; Galbo, H.: Effect of physical training on transport and metabolism of glucose in adipocytes. Am. J. Physiol. *244:* E129–E134 (1983).
112 Wahrenberg, H.; Engfeldt, P.; Arner, P.; Wennlund, A.; Östmann, J.: Adrenergic regulation of lipolysis in human adipocytes; Findings in hyper- and hypothyroidism. Clin. Endocrinol. Metab. *63:* 631–638 (1986).

113 Wardzala, L.J.; Crettaz, M.; Horton, E.D.; Jeanrenaud, B.; Horton, E.S.: Physical training of lean and genetically obese Zucker rats. Effect on fat cell metabolism. Am. J. Physiol. *243:* E418–E426 (1982).

114 Williams, R.S.; Bishop, T.: Enhanced receptor-cyclase coupling and augmented catecholamine stimulated lipolysis in exercising rats. Am. J. Physiol. *243:* E345–E351 (1982).

115 Winder, W.W.; Hickson, R.C.; Hagberg, J.M.; Ehsani, A.A.; McLane, J.A.: Training-induced changes in hormonal and metabolic responses to submaximal exercise. J. appl. Physiol. *46:* 766–771 (1979).

116 Wolfe, R.R.; Durkot, M.J.; Wolfe, M.H.: Investigation of kinetics of integrated metabolic response to adrenergic blockade in conscious dogs. Am. J. Physiol. *241:* E385–E395 (1981).

117 York, D.A.; Steinke, J.; Bray, G.A.: Hyperinsulinemia and insulin resistance in genetically obese rats. Metabolism *21:* 277–284 (1972).

Jens Bülow, MD, Department of Clinical Physiology/Nuclear Medicine,
Bispebjerg Hospital, DK–2400 Copenhagen NV (Denmark)

Poortmans JR (ed): Principles of Exercise Biochemistry.
Med Sport Sci. Basel, Karger, 1988, vol 27, pp 164–193.

Protein Metabolism

Jacques R. Poortmans

Chimie Physiologique, Institut Supérieur d'Education Physique et Kinésithérapie,
Université Libre de Bruxelles, Belgium

Introduction

The liver and, to a lesser extent the kidney, are the principal sites of amino acid metabolism in humans. When mammals are ingesting excess protein, amino acid amounts larger than needed for synthesis of proteins and other nitrogenous compounds cannot be stored or excreted and the surplus is oxidized or converted to carbohydrate and lipid. During amino acid degradation the α-amino group is removed and the resulting carbon skeleton is converted into a major metabolic intermediate. Most of the amino groups of amino acids are converted into urea by transdeamination, whereas their carbon skeletons are transformed into pyruvate, acetyl-CoA or one of the intermediates of the tricarboxylic acid cycle (fig. 1).

The loss of the α-amino group occurs by oxidative deamination (using the enzyme glutamate dehydrogenase) and transdeamination (using several aminotransferases and the glutamate dehydrogenase). Most of the amino acids can be converted to their respective oxoacids by aminotransferase (also called transaminase) reactions (fig. 2). All but two (lysine and threonine) amino acids appear able to be transaminated although it is not always clear how large a part these reactions play in the normal degradation of amino acids in the liver [75]. The reactions catalyzed at the amino-

Fig. 1. General pathway for the oxidation of amino acids.
Fig. 2. The transdeamination reactions and their importance to protein metabolism [75].

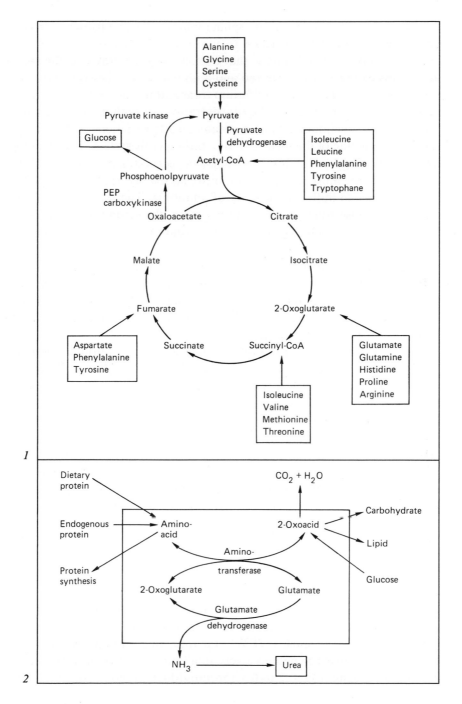

transferases (using pyridoxal phosphate-vitamin B6 as prosthetic group) and by glutamate dehydrogenase (using NAD$^+$ or NADP$^+$ as oxidizing agent) are close to equilibrium so that 2-oxoacids being provided, the overall process can be readily reversed and amino acids can be synthesized as well as degraded. The near-equilibrium transdeamination system provides an easy mechanism whereby the concentrations of both amino acids and 2-oxoacids are maintained fairly constant despite variations in the magnitude and direction of the metabolic flux through this system. The metabolism of amino acids, in addition to adenosine, generates most of the ammonia. Meanwhile, most tissues release nitrogen mainly as alanine or glutamine in order to buffer the toxicity of ammonia. The first reaction uses aminotransferase from glutamate to pyruvate, the second reaction transfers the ammonia itself to glutamate and is catalyzed by glutamine synthetase.

Although a large proportion of the ammonia does not arise from catabolism in the liver, the urea cycle occurs exclusively in the hepatic tissue and it requires four molecules of 'energy-rich' phosphate for the synthesis of one molecule of urea. In the human being as much as 90% of urinary nitrogen is in the form of urea.

The urea cycle appears to be regulated by nonequilibrium reactions with the first reaction being the flux-generating step. The synthesis of fumarate by the urea cycle is important because it links the urea cycle and the tricarboxylic acid cycle. In this respect, fumarate leading to oxaloacetate can be converted into glucose by a gluconeogenesis pathway.

Many cells are capable of concentrating amino acids from the extracellular environment, but prior to intracellular metabolism, amino acids must be transported across the cell membrane. There is strong evidence that a known set of reactions involving 5-glutamyl peptides (the γ-glutamyl transfer cycle of A. Meister) is used for amino acid transport into many tissues (kidney, small bowel, brain, erythrocytes). Another mechanism for the translocation of amino acids into cells involves at least seven different carriers.

The skeletal muscles, the intestines and the liver are particularly important in the disposal of excess amino acids. Much of the nitrogen is channeled into only a few compounds for the transport between the tissues (mainly alanine and glutamine). Free amino acid deposition in muscle often accounts for as much as 80% of the total amount in the whole body. In contrast, the plasma contains a very small proportion of the total amino acid pool, varying from 0.2 to 6% for individual amino acids. Thus, it is

Table I. Skeletal human muscle (70 kg body weight/40% muscle mass) [according to Bergström et al., 9, 10; Christman, 18]

Total intracellular free amino acid pool		86.5 g
Essential amino acids	8.4%	
Alanine	4.4%	
Glutamate	13.5%	
Glutamine	61.0%	
Taurine		34.6 g
Carnosine		1.4 g
Anserine		trace amount

not surprising that changes in the plasma amino acid levels should often reflect those in the free amino acid pool of the skeletal muscle.

This muscular pool in normal man weighing 70 kg has been calculated to be 86.5 g without taurine and 121.5 g with taurine [9]. Of the total pool of human skeletal muscle free amino acids, the eight essential amino acids represent only 8.4% whereas glutamine, glutamate and alanine constitute nearly 79% (table I). Among the amino acids the branched-chain amino acids (leucine, isoleucine and valine) are of particular interest as 60% of the total distribution of specific enzymes necessary for their oxidation (α-keto acid dehydrogenases) in man are located in skeletal muscle. Thus, it appears that muscle may be an important site for the catabolism of the branched-chain amino acids. These amino acids, unlike most of the others, are taken up by the striated muscles after a meal and partially oxidized in those tissues. The amino groups from all branched-chain amino acids can be reversibly transferred to 2-oxoglutarate and each of the remaining 2-oxocarboxylates then undergoes an irreversible oxidative decarboxylation which is catalyzed by the branched-chain keto acid dehydrogenase. In the postabsorptive period of starvation, the leg muscle of man releases essentially alanine and glutamine (60% of the total release). The origin of alanine is summarized in figure 3 [42] which shows that pyruvate derived from amino acids can be amined or oxidized. Glutamine is synthesized from glutamate and ammonia by the glutamine synthetase. The source of ammonia could arise from the glutamate dehydrogenase reaction as well as from the deamination of AMP by the purine nucleotide cycle. The endogenous synthesis of glutamate could arise from the combination of intermediates of the tricarboxylic acid cycle and two branched-chain amino acids (leucine and valine) as proposed by Newsholme and Leech [75].

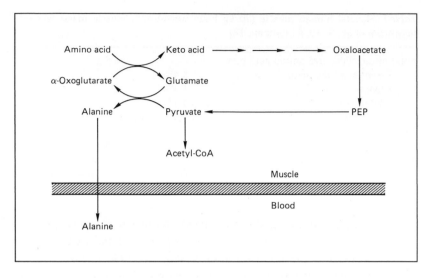

Fig. 3. Conversion of amino acids into pyruvate and alanine in skeletal muscle [42].

Amino acid oxidation in muscle leads to an appreciable amount of ATP generation. Assuming a common composition for the amino acids in muscle, the total balance of amino acid degradation is [66]:

1,000 mmol amino acids + 5,000 mmol O_2 → 695 mmol urea + 4,111 mmol CO_2 + 34 mmol SO_4^{2-} + 22,188 mmol ATP.

This equation also states that about 0.4 mol equivalent of glucose can be formed in the liver, or 72 g from the 110 g of mixed amino acids. The complete oxidation of leucine, isoleucine and valine gives 43, 42 and 32 mol of ATP, respectively, per amino acid molecule. However, the P/O ratio is only 2.4, compared to 2.8 for fats and 3.1 for glycogen, so amino acids are not a good fuel for maximum power production.

Short Historical Approach

The concept of protein and amino acid utilization during exercise has led to conflicting ideas. According to von Liebig [62], muscle protein breakdown was necessarily involved in muscular contraction. This opinion was contradicted by the work of Fick and Wislicenus [37], von Liebig's own students, who climbed the Faulhorn (1,956 m) on a nitrogen-free diet. From the 6 g of nitrogen excreted in the urine they calculated the protein used (i.e. 37.6 g) and thence the energy available on its combustion (i.e. 635 kJ), an amount far from adequate to supply the necessary energy for this event. Pettenkofer and Voit [78] confirmed this view, estimating that muscular work was without appreciable effect on nitrogen excretion. Since then, most authors have considered that protein is not used as a fuel during exercise when the caloric supply is adequate.

However, recent studies have demonstrated the effects of muscular exercises on nitrogen balance and whole-body protein metabolism in humans [23, 45, 79–81, 83, 88, 90]. Based on blood urea increase and nitrogen excretion, these studies have suggested that protein may in fact be degraded during exercise to an amount representing between 3 and 18% of the total energy expenditure. Meanwhile, amino acid oxidation seems to be negligible during exercise lasting for less than 1 h.

The present knowledge shows that when cellular homeostasis is interrupted by repeated bouts of a specific type of contractile activity the muscle responds by a selective adaptative change in the levels of certain proteins [14, 99].

The Glucose-Alanine Cycle

The effects of intense or prolonged exercise on plasma amino acid content has been studied in rats [19, 57] and in humans [17, 35, 55, 85]. Physical activity leads to a slight (10%) increase of total plasma amino nitrogen. However, the primary importance of alanine in amino acid outflow from peripheral depots is indicated by the fact that the arteriovenous difference for alanine is far in excess, in skeletal and heart muscle, of that of other amino acids. The transfer of alanine from muscle to blood increases with the intensity of exercise [84]. Since alanine represents only 11% of the free amino acid residues in skeletal muscle, peripheral synthesis of alanine by transamination of glucose-derived pyruvate has been suggested [34, 93]. A linear relationship exists between blood pyruvate and alanine. The K_m of the glutamate-pyruvate aminotransferase, which catalyzes the conversion of pyruvate into alanine, is near the intramuscular concentration of glutamate and pyruvate. The reversible reaction tends to reduce the metabolic acidosis which would appear if all pyruvate had been converted to lactate. Moreover, the α-amino nitrogen of the alanine released is largely derived from the catabolism of other amino acids. The synthesis of alanine may be related to the oxidation of amino acids, with alanine serving as a vehicle for the transfer of nitrogen in its nontoxic form from muscle to liver. On the basis of these observations, Felig and Wahren [35] have proposed the existence of a glucose-alanine cycle involving muscle and liver. The increased availability of pyruvate and amino acid groups in muscle, during exercise, results in augmented peripheral synthesis and greater release of alanine into the circulation. This excess of alanine is taken up by the liver and converted back to glucose which is released into the circulation and conducted to the muscle. Thus, alanine participates in gluconeogenesis in order to stabilize blood glucose.

The experiments of Odessey et al. [76] suggest that the release of branched chain amino acids by liver and muscle may be an important determinant of the rate of alanine synthesis by muscle (fig. 4).

Goldstein and Newsholme [42] estimate that the conversion of amino acid into pyruvate and alanine is not restricted to branched chain amino acids. Practically all amino acids could be converted to their keto acid analogues and thereafter to oxaloacetate and phosphoenolpyruvate. This means that a substantial energy supply could be provided from amino acids during exercise.

Felig and Wahren [35] have observed that the alanine release from skeletal muscle may increase from 25 µmol/min at rest to 175 µmol/min at 200 W. However, the splanchnic uptake of alanine only increases by 15–20% during moderate exercise and does not increase further during intense exercise. On the other hand, Hultman [50] has confirmed that, during prolonged light exercise, the total amino acid uptake by the splanchnic organs could at a maximum account for 8% of the glucose production and the alanine uptake for only half of that. Moreover, at high work loads less than 1% of the splanchnic glucose production could be attributed to alanine uptake.

Thus, the glucose-alanine cycle described by Felig and Wahren [35] is of quantitatively minor importance as a provider of energy substrate during exercise. Meanwhile, Wahren et al. [100] have observed significant net splanchnic uptake of alanine and other glycogenic amino acids during the recovery period, suggesting an accelerated rate of gluconeogenesis in the postexercise period. In their experiment these authors noted a figure of 9%, coming from amino acids, of the simultaneous glucose production as compared with the 6% in the basal state.

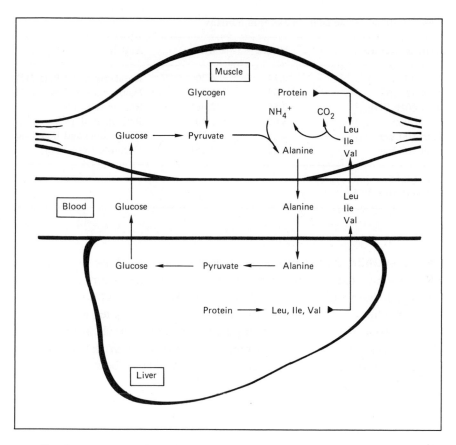

Fig. 4. Branched-chain-amino acid-alanine cycle between skeletal muscle and liver [76].

When the exercise exceeds 2 h in duration, a drop in plasma total amino acids and alanine occurs during the activity period [23]. As a consequence of this decrease in plasma content, amino acid concentration may go down to 60% of the pre-exercise level (table II). However, the resting conditions are obtained within 24 h after stopping the exercise. Berg and Keul [7] have established a relationship between the duration of the exercise and the alanine arteriovenous difference. From this study it appears that prolonged intense exercise partially depletes the amino acid pool. Under these conditions the alanine content no longer reflects the transamination of glucose-derived pyruvate.

Table II. Blood amino acids (mmol/l) in humans

Type of exercise	Before	After	Authors
Ski 12 km	4.19	4.20	Haralambie and Berg [45]
Run 42 km	4.08	2.90[1]	Haralambie and Berg [45]
Run 60% $VO_{2_{max}}$[2]	2.55	2.06[1]	Rennie et al. [90]
Ski 70 km	3.00	2.10[1]	Refsum et al. [87]
Run 100 km	4.26	2.87[1]	Haralambie and Berg [45]
Run 100 km	3.22	2.75[1]	Décombaz et al. [23]

[1] $p < 0.05$.
[2] Run during 3.75 h.

Meanwhile, during prolonged exercise, the splanchnic bed removes most amino acids during exercise and releases significant amounts of branched-chain amino acids. Ahlborg et al. [3] observed at 4 h of exercise that branched-chain amino acid removal by the working leg was balanced by an output of alanine, suggesting that these amino acids are providing the amino group for alanine formation. It is quite clear that changes in plasma amino acids during exercise represent a balance between the working muscle and the splanchnic area [44].

Amino Acid Catabolism

As mentioned in the introduction, catabolism of protein and amino acid leads to ammonia and urea production, oxidation of the carbon residues.

Ammonia Production

Ammonia is produced when muscle does work [64]. This production is proportional to the work done. Wilkerson et al. [105] have observed a significant exponential relationship between blood ammonia levels and lactate during exercise in humans. The immediate source of ammonia is the deamination of adenosine monophosphate (AMP) within the muscle, by adenylate deaminase, leading to the formation of inosine monophosphate (IMP). This reaction is an irreversible one but the regeneration of AMP from IMP occurs in two reactions which are catalyzed by adenylosuccinate synthetase and adenylosuccinase. Skeletal muscle can produce ammonia from aspartate. The conversion of aspartate to fumarate and ammonia occurs via a cyclic process termed the purine nucleotide cycle [64]. Experimental results provide evidence for the operation of the purine nucleotide cycle in skeletal muscle under conditions that are associated with an increased rate of glycolysis, such as exercise [43]. Ammonia production is greatest during intense exercise, when the rate of ATP utilization may exceed the rate of ATP formation. The myokinase reaction – 2-adenosine disphosphate (ADP) $\rightarrow$ ATP + AMP – leads to AMP accumulation. The removal of AMP by deamination stabilizes the relative ratios of ATP to ADP and AMP by favoring the production of ATP through the myokinase reaction. The deamination of AMP may contribute to the control of glycolysis and glycogenolysis by controlling a high ATP/ADP ratio and by supplying ammonium ion, an activator of phosphofructokinase.

Meyer and Terjung [69] have reported that AMP deamination occurs readily in rat fast-twitch glycolytic and fast-twitch oxidative glycolytic fibers, but not in slow-twitch oxidative fibers during stimulation in situ. Treadmill running of rats indicates that the fast-twitch muscle fibers, and, particularly, the glycolytic fibers, are the source of the ammonia produced during strenuous exercise [68]. It has been proposed that the purine nucleotide cycle functions together with amino acid deamination in muscle [43]. This hypothesis might be important in working muscle, since the fumarate produced could supply the Kreb's cycle during aerobic metabolism. Moreover, the IMP reamination must take place in order to restore the adenine nucleotide pool and this process requires α-amino groups from aspartate. By blocking IMP reamination with hadacidin (N-formyl-N-hydroxyaminoacetate), a competitive inhibitor of aspartate binding to adenylosuccinate synthetase, Meyer and Terjung [70] have provided evidence that this process occurs primarily during recovery after muscle use. Therefore, it is unlikely that AMP reamination functions as a significant pathway for amino acid deamination during work.

Urea Production

When short-term intense exercise is considered, blood urea remains relatively stable. Several authors have observed that the urea levels increase during long-lasting exercise, beyond 60–70 min [8, 23, 45, 81, 87, 88]. When expressed on a semilogarithmic scale, a close relationship appears between the increase in serum urea and the exercise duration (fig. 5). In addition, Haralambie and Berg [45] have shown a significant corre-

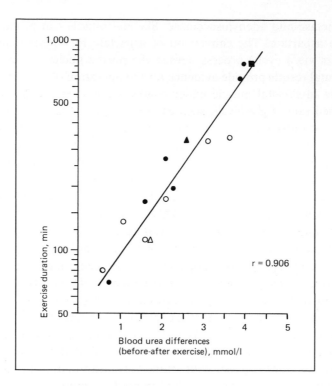

Fig. 5. Blood urea level and duration of running in humans. Each dot represents the mean value of various running events of different lengths conducted during performance conditions. ● = Haralambie and Berg [45]; ▲ = Refsum et al. [87]; ■ = Décombaz et al. [23]; ○ = Berg and Keul [8]; △ = Poortmans [81].

lation (r = −0.905) in serum between the increase of urea and the decrease of α-amino nitrogen. These close relationships suggest an enhanced protein (amino acid) oxidation during prolonged exercise. Table III presents data on urea production collected from the literature [23, 30, 58, 81, 87, 90]. The equation for the complete oxidation of amino acids to CO_2 and water permits, theoretically, an estimation of the total amount of protein oxidation during the exercise. It can be seen that between 3.8 and 9.6 g of proteins are oxidized per hour of work. Assuming a basal resting value of 1–2 g/h, the additional amino acid degradation suggests that protein may contribute significantly to the total expenditure. However, the energy utilization above resting contributed by amino acid degradation represents only 3–6% of the total energy cost.

Table III. Blood urea (mmol/l) in humans

Type of exercise	Before	After	Authors
Ski 12 km	7.74	8.46[1]	Haralambie and Berg [45]
Run 20 km	5.17	6.39[1]	Poortmans [81]
Walk 50 km	6.79	8.86[1]	Haralambie and Berg [45]
Ski 70 km	5.50	8.10[1]	Refsum et al. [87]
Ski 90 km	8.75	13.75[1]	Refsum and Strömme [88]
Run 100 km	7.70	11.56[1]	Haralambie and Berg [45]
Run 100 km	7.50	11.33[1]	Décombaz et al. [23]

[1] $p < 0.05$.

The extra protein oxidation could make some contribution to gluco-neogenesis, knowing that 1 g of amino acid gives an average of 0.65 g of glucose. According to table III, the protein-derived gluconeogenesis could account for a mean glucose production to 3.8 g/h. This amount is not negligible but far below the muscle glucose uptake during intense exercise.

The estimated amino acid degradation is mainly based on blood level and urine excretion of urea nitrogen. However, it is known that the sweating mechanism in man is an important mode of urea nitrogen excretion during exercise. Therefore, table III probably underestimates the magnitude of protein catabolism during exercise. The results obtained by Lemon and Mullin [58] demonstrate the importance of this compartment.

Moreover, the effect of availability of muscle glycogen upon protein metabolism has recently been observed in exercising humans [58]. After carbohydrate depletion, sweat urea excretion was equivalent to a protein breakdown of 13.7 g/h or 10.4% of the total calorie cost, while after carbohydrate loading the protein degradation was restricted to 5.8 g/h (4.4% of total calorie cost). This fact suggests that the initial muscle glycogen levels may be a regulating factor in exercise protein catabolism.

As well, raised plasma free fatty acid levels induced by lipid infusion have a hypoaminoacidemic action in resting humans [36]. Since prolonged exercise of medium intensity raises the free fatty acid level in plasma, it may be argued that this effect could lower the plasma amino acid level by some unresolved mechanism (decrease in the rate or release or increase in the rate of disappearance).

Eventually, the indirect estimation of changes in urea production by measuring excreted urea, and estimating sweat loss and changes in urea loss pool size, has many potential pitfalls. Indeed, despite the increase in amino acid oxidation, Wolfe et al. [107] did not observe changes in urea concentration and production.

Amino Acid Oxidation

Since a few years the loss of total protein during endurance exercise has been reported in rat gastrocnemius [29]. Direct determination on exercising muscle samples shows a decrease in alanine, glutamine and glutamate, whereas the levels of aspartate and the branched-chain amino acids are higher [19, 24, 106]. Previous papers report that exercise results in the loss of liver total protein [24, 29, 52].

A limited degradation (11–14%) of liver and muscle proteins has been observed during exhaustive treadmill running in rats [29, 52]. This protein loss in liver is quite sizeable, amounting to about 0.23 g/h for a 300-gram rat; the caloric value of the liver protein lost could account for as much as 19% of the energy utilized during exercise, if all amino acids had been oxidized. This protein degradation was accompanied in liver [52, 53] and in muscle [27] by an increase in the activities of free cathepsin D, suggesting that lysosomal enzymes might be involved in the process.

All free amino acids but glutamine are enhanced from 30 to 300% in liver tissue under the influence of endurance exercise [24]. Thus, it seems that the liver is providing free amino acids to the working muscle.

The influence of short-term [19] and long-term [24] swimming exercise on free amino acid distribution in muscle and liver has been investigated in rats. In both cases, glutamine and glutamic acid levels were depressed while aspartate concentration was enhanced in muscle. Exercise caused a lowering in glycine and an increased level of serine in liver. The branched chain amino acids were generally elevated by exercise as were lysine and tyrosine in muscle and liver. The changes observed in the amino acid contents of muscle and liver are consistent with the increase of protein degradation during exercise.

The uptake by the muscle of arterial amino acids has been investigated during electrical stimulation of isolated frog sartorius [74] and rat diaphragm [41] using α-aminoisobutyrate, a model substrate employed in the study of amino acid transport. These various experiments indicate that muscular work provokes an accumulation of the amino acid analogue and that the effects of stimulation on the transport of the amino acid persisted undiminished for several hours after ceasing the treatment [41, 74]. Further insight into amino acid use has come from an arteriovenous difference across the exercising leg and the splanchnic vascular bed in man. The works of Felig and Wahren [35] and Ahlborg et al. [3] have shown a significant splanchnic output of valine, leucine and isoleucine after 4 h of mild exercise. These amino acids are taken up by the exercising muscle.

Stable isotope probes have been used to investigate the changes in amino acid oxidation that accompany exercise. Several authors [4, 25, 32, 60, 90, 104, 107] have demonstrated in vitro and in vivo that the oxidative decarboxylation of several amino acids is enhanced during physical exercise. Marked estimation by exercise of skeletal muscle branched-chain oxoacid dehydrogenase observed by Kasperek and Snider [53, 54] may be initiated by increasing leucine concentration [2]. Most investigators chose to quantify oxidation of the essential amino acid leucine, $(1-^{13}C)$, $(1-^{14}C)$ or $(U-^{14}C)$, as an index of net protein catabolism. Their data demonstrate that the rate of leucine oxidation is proportionably related to the intensity of exercise in rats [60, 104] and in humans [72]. During exercise, leucine oxidation rates increase by 2.4- to 30-fold that of the resting level. This increased quantity of oxidized leucine suggests that prolonged exercise may alter the subject's daily requirement for amino acid if repeated daily. Evans et al. [32] estimate that nearly 900 mg of leucine is oxidized during a 2-hour ride on a cycle ergometer at 55% of VO_{2max}. This represents almost 90% of the current estimated leucine requirement! The precise origin and site of the increased leucine oxidation is difficult to determine. Meanwhile, Wolfe et al. [107] believe that, in exercising humans, about half of the leucine oxidized comes from plasma and the remainder from intracellular pools, mainly muscle. Figure 6 summarizes our perceptions drawn from the existing literature, illustrating the response of proteins and amino acid utilization to a single bout of endurance exercise.

Quite recently, the use of specific activity of leucine for the calculation of the rate of oxidation of amino acids has been questioned. Wolfe et al. [108] observed during exercise an increase oxidation smaller for lysine than for leucine, with urea production being unaffected. Moreover, it appears that leucine or 2-oxoisocaproate specific activity in muscle is better predicted by plasma 2-oxoisocaproate than leucine-specific activity [98]. These results emphasize the need to test more than one amino acid to measure whole-body protein catabolism.

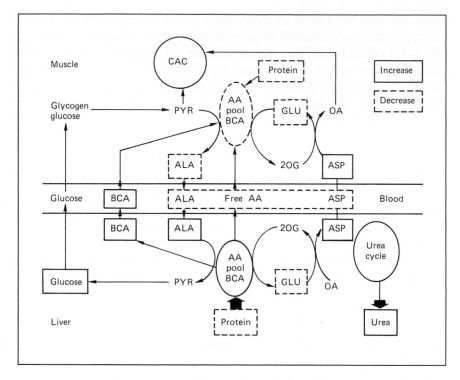

Fig. 6. Amino acid catabolism during endurance exercise. CAC = Citric acid cycle;
AA = amino acid; BCA = branched-chain amino acid; 2OG = 2-oxoglutarate.

Renal Handling of Amino Acids

There are relatively few studies on urinary amino acid excretion during physical activities. Meanwhile, this compartment must be taken into consideration when evaluating the total amino acid balance. Therefore, we have investigated plasma and urine amino acid changes, and thereby their individual renal clearances, in a 100-km run [23]. The rate of total urinary excretion of 22 amino acids decreases from 4.3 μmol/min before the run to 1.6 μmol/min after the run. The clearance of most amino acids was reduced to 2- to 3-fold during the run. The amino acids efficiently reabsorbed at rest exhibited a less-reduced clearance immediately after the run, as compared to the amino acids with poor resting tubular reabsorption. Assuming an average duration of 700 min for the ultramarathon, this means that the renal loss of amino acid is reduced from 0.30 g at rest to 0.12 g after exercise. Therefore, the urinary excretion of amino acids during exercise has a minor effect on the total balance of these fuels.

Protein Turnover during and after Exercise

General Aspects

The term turnover covers both the synthesis and breakdown of protein. In the steady state condition the energy cost of protein synthesis will approximately account for 10% of the basal oxygen uptake [102]. Total protein synthesis in human adult subjects is about 3.0 g/kg/day [101] while protein turnover is about 5.7 g/kg/day [102]. Protein degradation in human skeletal muscles estimated from the release of tyrosine in the presence of insulin and amino acids is approximately 34 nmol/h/g wet weight. This degradation rate corresponds to a half-life of approximately 20 days [65].

Already in 1895, Morpurgo found that strenuous exercise could induce the cross-sectional area of the sartorius muscle to increase without altering the total number of muscle fibers. This muscular hypertrophy following exercise has been repeatedly confirmed. Several reviews have recently been focused on that subject [13, 44, 56, 59, 79–81, 83].

It has been shown that electrical stimulation enhances amino acid uptake by the muscle [41, 74]. The uptake depends on both the intensity and duration of the stimulation. It is important to note that the effects of electrical stimulation persisted essentially undiminished for 5–6 h after cessation of the treatment. However, it is essential to distinguish between an increase in amino acid uptake and protein synthesis. In normal resting humans, the response to an influx of amino acids seems to be storage of proteins, which is broken down again as soon as the inflow ceases [102]. The situation appears to be completely different during exercise. On the contrary, no stimulation of amino acid incorporation into proteins has been observed but rather a depression on protein synthesis in heart [20, 96] and skeletal muscle [5, 27, 89, 90, 103]. This inhibition may rise to 35–55% according to the cellular compartment and the tissue involved. Rennie et al. [89] have pointed out that at the end of 40 min of exhaustive exercise a reduction of actin synthesis down to 23% of the basal level occurs. Prolonged exercise also alters the rate of protein degradation in perfused rat muscle [27] and in human whole-body protein [90]. Exercise

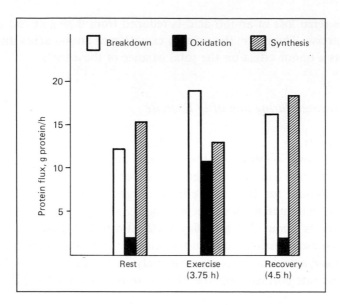

Fig. 7. Protein metabolism in humans (74 kg body weight) during and after strenuous exercise [90].

causes a significant increase in protein degradation. Together with the fall in whole-body protein synthesis, this may provide an explanation for the usually observed negative nitrogen balance associated with exercise. However, in muscle it is certainly important to distinguish between the effects of passive stretch and active contraction; Goldberg et al. [41] have shown that the rate of protein degradation depends upon the level of muscular activity. Subsequently, passive stretch and active contraction appear to have additive effects in delaying breakdown. After exercise a positive whole-body nitrogen balance is regained by an increase in whole-body synthesis over whole-body protein breakdown (fig. 7). The increase in protein synthesis remains for several hours following the end of exercise [20, 90, 96, 103]. In vitro studies with guinea pig skeletal muscle suggest that the increase of amino acid incorporation into proteins is possibly due to an increase in ribosomal translational activity [92]. Indeed, the polysome profiles of controls and exercised muscle are identical and no difference in the total polysome RNA content could be detected. These observations could be linked with the observed increase in amino acid uptake by skeletal muscle during the recovery period.

Recently, Booth and Watson [14] have summarized the in vivo and in vitro information on the directional change in protein synthesis rate of skeletal muscle during exercise. Increased contractile activity of less than 30 min is associated with decreased muscle protein synthesis rates of skeletal muscle. After short bouts of exercise, protein synthesis rates in muscles appear to decrease in the first hour after exercise, but increase in the second hour after exercise to levels higher than normal. Kasperek and Snider [53] have shown that the increased rate of protein degradation affected both red and white muscles during downgrade running rats.

3-Methylhistidine Measurements

The measurement of the amino acid 3-methylhistidine (3-MeHis) in urine has been recognized as a useful, noninvasive technique for assessing muscle protein breakdown. The major site of endogenous 3-MeHis is the actin of all muscle fibres and the myosin of white muscle fibres. During catabolism of these myofibrillar proteins, the released 3-MeHis is neither reused for protein synthesis nor metabolized oxidatively. Therefore, the total urinary excretion of 3-MeHis has been suggested as an in vivo index of muscle protein breakdown in man [110]. There are conflicting reports on the quantitative importance of nonskeletal muscle 3-MeHis in rat urine, i.e. from 10 to 75% [46, 71]. Data of Aftring et al. [1] have shown that skeletal muscle tissue is the major contributor of 3-MeHis in human urine, since it contributes as much as 75% of the urinary excretion.

However, Rennie and Millward [91] still believe that the urinary 3-MeHis is not a valid index of skeletal muscle production of 3-MeHis. They give evidence that the use of 3-MeHis excretion or of the urinary 3-MeHis/creatinine ratio as indices of skeletal protein breakdown should be discontinued. However, they believe that the measurement of arteriovenous differences of 3-MeHis across skeletal muscle could be used to provide useful information about the regulation of myofibrillar protein breakdown.

Décombaz et al. [23] were the first to determine the level of 3-MeHis in plasma and urine during a 100-km run. Since then a few studies have added information to the princeps movens investigation. Table IV reports the results obtained in different events in men. In all studies but two, the plasma concentration of 3-MeHis seems to remain stable during prolonged exercise, while its urinary excretion decreased in most studies. Even after creatinine correction the urine elimination of 3-MeHis is depressed. The level of this free amino acid in muscle is reduced by 30% during 40 min of

Table IV. Plasma level and urinary excretion of 3-MeHis (humans) – means

Type of exercise	Duration min	Plasma, μmol/l		Urine excretion				Authors
				nmol/min		μmol/mmol creatinine		
		before	after	before	after	before	after	
Run	60	–	–	205	168[1]	24.0	21.6	Radha and Bessman [86]
Run 16 km	84	7.7	14.4	198	450[1]	29.4	36.2[1]	Dohm et al. [30]
Run 60% VO_{2max}	225	12.0	12.0	169	37[1]	17.7	9.2[1]	Rennie et al. [90]
Run 70% VO_{2max}	90	5.1	5.8[1]	–	–	19.7	14.9[1]	Dohm et al. [28]
Run 100 km	750	23.4	26.0	351	182[1]	27.6	21.0[1]	Décombaz et al. [23]

[1] $p < 0.05$.

exercise [90]. However, calculation in men undergoing a 100-km run shows that the total production of 3-MeHis remains unchanged [23].

Some disparate results have been explained by results of an experiment by Dohm et al. [26] who followed the 3-MeHis excretion in humans during successive hours after exercise. These authors observed a biphasic change of 3-MeHis excretion in response to exercise with an immediate decrease in the 3-MeHis-to-creatinine ratio during exercise followed by a prolonged increase for several hours.

Buffering Capacity of Amino Acid-Related Substances

During exhaustive exercise, hydrogen ions are released within the muscle. According to Hultman and Sahlin [51] nearly 94% of the protons are the result of accumulation of lactic acid; the remaining are due to the presence of other acids (pyruvate 0.3%; malate 3%) and to accumulation of glucose 6-phosphate (2%) and glycerol-1-phosphate (1%). The resting muscle pH of 7.0 will therefore decrease to about 6.6 after severe exercise and the concentration of the released H^+ ions is thus taken up by different buffering processes.

The buffering capacity in muscle, or the extent of pH decrease after addition of H^+ ions, may be divided into three major components: physicochemical buffering, consumption and production of nonvolatile acids, transmembrane fluxes of H^+ and HCO_3^- [95].

Skeletal muscle contains large amounts of amino acids, peptides and proteins with fully ionized α-carbonyl and α-amino groups at pH 7.0. The buffering capacity of these compounds depends on the acid-base behavior of their R groups. Only histidine has a pKa value of its R group which is the physiological pH range. Other imidazole-containing dipeptides with pKa between 6.8 and 7.0 are found in mammalian skeletal muscle, namely carnosine (β-alanyl-*L*-histidine) and anserine (β-alanyl-*L*-1-methylhistidine). Eventually, protein-bound histidine residues contribute to the buffer capacity in human muscle. Quantitative data have shown that the buffer value of human muscle mainly depends on its protein content (46 mmol/l of histidine) [39], the concentration of carnosine being 6.5 mmol/l [10], and that of anserine being insignificant in man [18]. On the basis of the histidine content, Hultman and Sahlin [51] have calculated the minimum uptake of H^+ ions by protein-related substances. The buffering capacity of these compounds amounts to 12.5 mmol/l when the muscle pH changes from 7.08 to 6.6 after 5–10 min exhaustive bicycle exercise. Added to the 1.8 mmol/l uptake of H^+ to carnosine, the total uptake of H^+ ions in exercising muscle due to physicochemical buffering of protein-related substances amounts to 14.3 mmol/l or 57% of the whole physicochemical processes. The importance of buffering the accumulated protons is obvious, since decrements in pH may affect exercise performance. A high concentration of taurine (table I) is present in skeletal muscle amounting to nearly 75% of total body taurine. The significance of the high level of taurine in muscle is not yet known. Very recently, Hitzig et al. [48] have determined brain amino acid concentrations during diving and acid-base stress in turtles. They found that taurine decreased significantly in the diving animals and increased significantly in those subjected to anoxia plus hypercapnia. The authors suggest that taurine could act as neurohormones confined to particular areas within the central nervous system.

When protein and amino acids are oxidized, the net reaction of deamination and oxidation of the formed keto acid results in a net uptake of protons. During short-term exercise Bergström et al. [11] have recorded an increase in muscle glutamine of 4.2 mmol/l and an inosine monophosphate production of 1.1 mmol/l. The net uptake of protons during this exhaustive exercise due to amino acid catabolism is thus about 3.1 mmol/l. Thus, glutamic acid is contributing to a small extent to the buffer capacity of skeletal muscle.

Control Adaptations in Protein Levels

Probably the best evidence of distinct adaptive changes in the level of skeletal protein is the loss of muscle mass under conditions in which force production and contraction are reduced. Inactivity or immobilization of limbs [12, 63] or legs [40] induces a consequent loss of muscle force and mass. There is no evidence for a change in the number of muscle fibres per unit cross-sectional area, but there is significant decrease in muscle fibre diameter. The results obtained in human quadriceps muscle show that the decrement in muscle mass is mainly due to a substantial (25%) depression of muscle protein synthesis [40].

Mechanical and chemical factors have been recognized to produce muscle hypertrophy [41]. Besides the stretching of the sarcolemna, the control of muscle protein synthesis could be modulated, at the translational level, by the cytoplasmic redox potential.

In vitro studies with isolated diaphragms suggest that increasing the $NAD^+/NADH$ ratio has been associated with accelerated protein degradation, stimulation of branched chain amino acid oxidation and alanine production. In well-perfused skeletal muscle NAD becomes more oxidized during exercise. Muscle protein synthesis is partially inhibited. Thus, the conditions suggested by the model of Hedden and Buse [47] seem to prevail during submaximal exercise. An increase in the cystolic $NAD^+/NADH$ ratio is accompanied by a decrease in protein synthesis.

Along the same lines Bylund-Fellenius et al. [16] showed reduced 3-MeHis concentration and release in association with increased lactate output during exercise. Also, Tischler and Fagan [97] reported correlations between the redox state (NAD, NADP) and proteolysis, showing that increased proteolysis is associated with a more oxidized state. All these data suggest that a depletion of high-energy phosphates and accumulation of lactate could inhibit degradation on contracting skeletal muscle. An increased ratio of NAD (P) H/NAD (P)$^+$ in muscle during short-term exercise may decrease degradation.

Booth and Watson [14] postulate changes in mRNA translation and peptide elongation as possible regulation of the rates of protein synthesis in skeletal muscles. However, no single model is applicable to all types of exercise, nor is any method completely free of assumptions and limitations. This is why, at the present, it remains difficult to present a well-balanced view on the mechanisms controlling protein level in exercise.

Influence of Endurance Training

Endurance training induces an adaptive increase in the mitochondrial content and respiratory capacity of skeletal muscle [49]. Meanwhile, a few studies only have been devoted to protein metabolism adaptations. Dohm et al. [25] have demonstrated that amino acid catabolism is increased during exercise training and that the muscle enzymes involved in leucine oxidation adapt to endurance training in a manner similar to the enzymes of carbohydrate and fat oxidation. Since the increased urea excretion suggested increased amino acid oxidation in endurance-trained rats, these authors have also reported an enhanced rate of $^{14}CO_2$ production from [1-^{14}C]leucine in vivo. The influence of endurance training on the rate of uptake and levels of free amino acids in plasma and various tissues has been investigated in rats [6]. Training decreased the uptake of α-aminoisobutyrate (an amino acid analogue) into the heart but not into muscle and liver. Lastly, Molé et al. [73] have looked at the adaptations to exercise of the enzymatic pathways of pyruvate metabolism in rat skeletal muscle. They reported that the level of activity of glutamate-pyruvate aminotransferase was increased approximately 85% in both the mitochondrial and cytoplasmic fractions of gastrocnemius muscles of the runners.

It seems possible that this adaptation could result in the conversion of a greater percentage of the pyruvate formed in the muscle during exercise to alanine and less to lactate by increasing the capacity of glutamate-pyruvate aminotransferase to compete with lactate dehydrogenase for pyruvate. In the same line of evidence, the level of activity of mitochondrial aspartate-oxoglutarate aminotransferase increased 2-fold in leg muscles of rats subjected to the programme of running [73].

Dohm et al. [26] have reported that daily performance of exercise may have a cumulative effect on 3-MeHis, suggesting that athletes on a strenuous training schedule may have increased muscle protein turnover. However, experiments on endurance-trained rats do not show any affect on basal rates of protein synthesis and degradation in muscle [21, 22]. More-over, prior exercise training has little effect on the response of muscle protein turnover after acute exercise.

A few authors have investigated the buffering capacity of human mus-cle in trained and untrained subjects. As compared to sedentary subjects a higher buffer capacity was observed in muscle of trained subjects in sports where a high degree of anaerobic energy utilization is required [77, 94]. Sprint-trained athletes have a higher buffer capacity than marathon run-ners [77]. A higher concentration of carnosine was observed in the vastus lateralis muscle of sprinters and rowers. Meanwhile, the contribution of carnosine to maximal buffer capacity could contribute to at most 10% of total buffering. Thus, the observed increase in carnosine levels may only partially contribute to the enhanced buffering capabilities of the sprint-trained subjects.

Muscle protein appears to be the most important intracellular buffer. According to Sahlin and Henriksson [94], it is therefore conceivable that the difference in buffer capacity is mainly due to a difference in the protein composition of trained and untrained subjects.

Protein Requirements for Athletes

According to the Food and Nutrition Board and the FAO/WHO [33], the minimum recommended daily allowance is 0.9 g of protein/kg BW in adults. All agree that adolescents should have an additional supply of nitrogen requirements, provided that the distribution of essential amino acid content is optimal. Also, the nutritional value of ingested protein must be taken into consideration knowing that the biological value (retained nitrogen/absorbed nitrogen) would be 1.00 if all the amino acids could be used for synthesis of protein. Cows' milk protein has a biological value of 0.95 while that of whole corn has a biological value of 0.60 [75]. Measure-ments of whole body amino acid kinetics with the aid of ^{13}C-tracers led Meredith et al. [67] to postulate that the FAO/WHO recommendations, derived from results of short-term N-balance determinations, are too low to the extent of a factor of about 2–3. Thus, the complete and accurate assessments of the quantitative significance of indispensable amino acids in the protein nutrition of healthy adults at rest is still lacking.

Fuge et al. [38] have evaluated the effects of a protein-deficient diet on some of the enzymatic adaptive changes that occur in rats in response to strenuous endurance exercise. Their results indicated that protein deficiency did not prevent the increase in the capacity of gastrocnemius muscles of the exercising rats to oxidize pyruvate. Moreover, the protein-deficient rats were able to run significantly longer than the runners on a normal diet. Therefore, in rats, an overfeeding with proteins does not seem to have a beneficial effect on the endurance performance.

Contradictory results have been obtained in human subjects submitted to muscular activities. Earlier literature believes that protein requirements for athletes should be restricted to the minimum recommended daily allowance [31], while more recent studies claim that athletes should consume 1.8–2.0 g of protein/kg BW per day, or twice the requirement for sedentary individuals [15, 61]. Knowing that most athletes oxidize about 5 g protein hourly, in addition to their basal value [81], it looks wise to suggest a reasonable amount of 1.2 g of protein/kg BW per day for subjects engaged in a 2-hour endurance or strength/power daily program. It must be an equilibrated basal diet and protection should be taken towards undetected minor protein deficiencies which could occur in growing adolescents and athletes oriented towards excess carbohydrates.

References

1 Aftring, E.G.; Bernhard, W.; Janzen, R.W.C.; Röthig, H.J.: Quantitative importance of non-skeletal muscle N-methylhistidine and creatine in human urine. Biochem. J. 200: 449–452 (1981).

2 Aftring, R.P.; Block, K.P.; Buse, M.G.: Leucine and isoleucine activate skeletal muscle branched-chain α-keto acid dehydrogenase in vivo. Am. J. Physiol. 250: E599–E604 (1986).

3 Ahlborg, G.P.; Felig, P.; Hagenfeldt, L.; Hendler, R.; Wahren, J.: Substrate turnover during prolonged exercise in man. Splanchnic and leg metabolism of glucose, free fatty acids and amino acids. J. clin. Invest. 53: 1080–1090 (1974).

4 Askew, E.W.; Klain, G.F.; Lowder, J.F., Jr.; Wise, W.R., Jr.: Influence of exercise on amino acid mobilization and oxidation. Med. Sci. Sports 11: 106 (1970).

5 Bates, P.C.; De Coster, T.; Grimble, G.K.; Holloszy, J.O.; Millward, D.J.; Rennie, P.J.: J. Physiol., Lond. 303: 41P (1980).

6 Beecher, G.R.; Puente, F.R.; Dohm, G.L.: Amino acid uptake and levels: influence of endurance training. Biochem. Med. 21: 196–201 (1979).

7 Berg, A.; Keul, J.: Serum alanine during long-lasting exercise. Int. J. Sportmed. 1: 199–202 (1980).

8 Berg, A.; Keul, J.: Physiological and metabolic responses of female athletes during laboratory and field exercise; in Borms, Hebbelinck, Venerando, Women and sport, pp. 77–96 (Karger, Basel 1981).

9   Bergström, J.; Fürst, P.; Norée, L.O.; Vinars, E.: Intracellular free amino acid con-
    centration in human muscle tissue. J. appl. Physiol. *36:* 693–697 (1974).
10 Bergström, J.; Fürst, P.; Norée, L.O.; Vinnars, E.: Intracellular free amino acids in
    muscle tissue of patients with chronic uraemia: effect of peritoneal dialysis and
    infusion of essential amino acids. Clin. Sci. mol. Med. *54:* 51–60 (1978).
11 Bergström, J.: In Hultman, Sahlin, Acid-base balance during exercise. Ex. Sports Sci.
    Rev. *8:* 102–103 (1980).
12 Booth, F.W.; Butler, D.T.; Nicholson, W.F.; Watson, P.A.: Insulin resistance for
    protein synthesis does not occur in muscles of immobilized rat hindlimb. Med. Sci.
    Sports *14:* 108 (1982).
13 Booth, F.W.; Nicholson, W.F.; Watson, P.A.: Influence of muscle use on protein
    synthesis and degradation. Ex. Sport Sci. Rev. *10:* 27–48 (1982).
14 Booth, F.W.; Watson, P.A.: Control of adaptations in protein levels in response to
    exercise. Fed. Proc. *44:* 2293–2300 (1987).
15 Brotherhood, J.R.: Nutrition and sports performance. Sports Med. *1:* 350–389
    (1984).
16 Bylund-Fellenius, A.C.; Ojama, K.M.; Flaim, K.E.; Li, J.B.; Wassner, S.J.; Jefferson,
    L.S.: Protein synthesis versus energy state in contracting muscles of perfused rat
    hindlimb. Am. J. Physiol. *246:* E297–E305 (1984).
17 Carlson, A.; Hallgren, B.; Jagenburg, R.; Svanborg, A.; Werko, L.: Arterial concen-
    trations of free fatty acids and free amino acids in healthy human individuals at rest
    and at different work loads. Scand. J. clin. Lab. Invest. *14:* 185–191 (1962).
18 Christman, A.A.: Factors affecting anserine and carnosine levels in skeletal muscles
    of various animals. Int. J. Biochem. *7:* 519–527 (1976).
19 Christophe, J.; Winand, J.; Kutzner, R.; Hebbelinck, M.: Amino acid levels in plas-
    ma, liver, muscle and kidney during and after exercise in fasted and fed rats. Am. J.
    Physiol. *221:* 453–457 (1971).
20 Cook, E.A.; Taylor, P.B.; Swartman, J.R.: Effect of acute exercise on amino acid
    incorporation into the rat myocardium. Eur. J. appl. Physiol. *47:* 105–111 (1981).
21 Davis, T.A.; Karl, I.E.; Tegtmeyer, E.D.; Osborne, D.F.; Klahr, S.; Harter, H.R.:
    Muscle protein turnover: effects of exercise training and renal insufficiency. Am. J.
    Physiol. *248:* E337–E345 (1985).
22 Davis, T.A.; Karl, I.E.: Response of muscle protein turnover to insulin after acute
    exercise and training. Biochem. J. *240:* 651–657 (1986).
23 Décombaz, J.; Reinhardt, P.; Anantharaman, K.; Glutz, G. von; Poortmans, J.: Bio-
    chemical changes in a 100-km run. Free amino acids urea and creatinine. Eur. J.
    appl. Physiol. *41:* 61–72 (1979).
24 Dohm, G.L.; Beecher, G.R.; Warren, R.Q.; Williams, R.T.: Influence on exercise on
    free amino acid concentrations in rat tissues. J. appl. Physiol. *50:* 41–44 (1981).
25 Dohm, G.L.; Hecker, A.L.; Brown, W.E.; Klain, G.J.; Puente, F.R.; Askew, E.W.;
    Beecher, G.R.: Adaptation of protein metabolism to endurance training. Biochem. J.
    164: 705–708 (1977).
26 Dohm, G.L.; Israel, R.G.; Breedlove, R.L.; Williams, R.T.; Askew, E.W.: Biphasic
    changes in 3-methylhistidine excretion in humans after exercise. Am. J. Physiol.
    248: E588–E592 (1985).
27 Dohm, G.L.; Kasperek, G.J.; Tapscott, E.B.; Beecher, G.R.: Effect of exercise on
    synthesis and degradation of muscle protein. Biochem. J. *188:* 255–262 (1980).

28 Dohm, G.L.; Kasperek, G.J.; Tapscott, E.B.; Barakat, H.A.: Protein metabolism during endurance exercise. Fed. Proc. *44:* 348–352 (1985).

29 Dohm, G.L.; Puente, F.R.; Smith, C.P.; Edge, A.: Changes in tissue protein levels as a result of endurance exercise. Life Sci. *23:* 845–850 (1978).

30 Dohm, G.L.; Williams, R.T.; Kasperek, G.J.; Van Rij, A.M.: Increased excretion of urea and N-methylhistidine by rats and humans after a bout of exercise. J. appl. Physiol. *52:* 27–33 (1982).

31 Durnin, J.V.G.A.: Protein requirements and physical activity; in Parizkova, Pogozkin, Nutrition, physical fitness and health, pp. 53–60 (University Park Press, Baltimore 1978).

32 Evans, W.J.; Fisher, E.C.; Hoerr, R.A.; Young, V.R.: Protein metabolism and endurance exercise. Phys. Sportsmed. *11:* 63–72 (1983).

33 FAO/WHO/UNU Meeting: Energy and protein requirements; technical report, No. 724 (WHO, Geneva 1986).

34 Felig, P.; Pozefsky, T.; Marliss, E.; Cahill, G.F., Jr.: Alanine, key role in gluconeogenesis. Science *167:* 1003–1004 (1970).

35 Felig, P.; Wahren, J.: Amino acid metabolism in exercising man. J. clin. Invest. *50:* 2703–2714 (1971).

36 Ferrannini, E.; Barrett, E.J.; Bevilacqua, S.; Jacob, R.; Walesky, M.; Sherwin, R.S.; De Fronzo, R.A.: Effect of free fatty acids on blood amino acid levels in humans. Am. J. Physiol. *250:* E686–E694 (1986).

37 Fick, A.; Wislicenus, J.: Über die Entstehung der Muskelkraft. Viertel-Jahrsschr. Zürcher Naturforsch. Ges. *10:* 317 (1986).

38 Fuge, K.W.; Grews, E.L., III; Pattengale, P.K.; Holloszy, J.O.; Shank, R.E.: Effects of protein deficiency on certain adaptive responses to exercise. Am. J. Physiol. *215:* 600–663 (1968).

39 Furst, P.; Jonsson, A.; Josephson, B.; Vinnars, E.: Distribution in muscle and liver vein protein of 15N administered as ammonium acetate to man. J. appl. Physiol. *29:* 307–312 (1970).

40 Gibson, J.N.A.; Halliday, D.; Morrison, W.L.; Stoward, P.J.; Hornsby, G.A.; Watt, P.W.; Murdoch, G.; Rennie, M.J.: Decrease in human quadriceps muscle protein turnover consequent upon leg immobilization. Clin. Sci. *72:* 503–509 (1987).

41 Goldberg, A.L.; Jablecki, C.; Li, J.B.: Effects of use and disuse on amino acid transport and protein turnover in muscle. Ann. N.Y. Acad. Sci. *228:* 190–201 (1974).

42 Goldstein, L.; Newsholme, E.A.: The formation of alanine from amino acids in diaphragm muscle of the rat. Biochem. J. *154:* 555–558 (1976).

43 Goodman, M.N.; Lowenstein, J.M.: The purine nucleotide cycle. Studies of ammonia production by skeletal muscle in situ and in perfused preparations. J. biol. Chem. *252:* 5054–5060 (1977).

44 Goodman, M.N.; Ruderman, N.B.: Influence of muscle use on amino acid metabolism. Ex. Sport Sci. Rev. *10:* 1–26 (1982).

45 Haralambie, G.; Berg, A.: Serum urea and amino nitrogen changes with exercise duration. Eur. J. appl. Physiol. *36:* 39–48 (1976).

46 Harris, I.: Reappraisal of the quantitative importance of non-skeletal muscle source of N-methylhistidine in urine. Biochem. J. *194:* 1011–1014 (1981).

47 Hedden, M.P.; Buse, M.G.: Effects of glucose, pyruvate, lactate and amino acids on muscle protein synthesis. Am. J. Physiol. *242:* E184–E192 (1982).

48   Hitzig, B.M.; Kneussl, M.P.; Shih, V.; Brandstetter, R.D.; Kazemi, H.: Brain amino
     acid concentrations during diving and acid-base stress in turtles. J. appl. Physiol. *58:*
     1751–1754 (1985).
49   Holloszy, J.O.; Rennie, M.J.; Hickson, R.C.; Conlee, R.K.; Hagberg, J.M.: Physio-
     logical consequences of the biochemical adaptations to endurance exercise. Ann.
     N.Y. Acad. Sci. *301:* 440–450 (1977).
50   Hultman, E.: Regulation of carbohydrate metabolism in the liver during rest and
     exercise with special reference to diet; in Landry, Orban, 3rd Int. Symp. Biochem-
     istry of Exercise, Miami, pp. 99–126 (Symposia Specialists, Miami 1978).
51   Hultman, E.; Sahlin, K.: Acid-base balance during exercise. Ex. Sport Sci. Rev. *8:*
     41–128 (1980).
52   Kasperek, G.J.; Dohm, G.L.; Tapscott, E.B.; Powell, T.: Effect of exercise on liver
     protein loss and lysosomal enzyme levels in fed and fasted rats. Proc. Soc. exp. Biol.
     Med. *164:* 430–434 (1980).
53   Kasperek, G.J.; Snider, R.D.: The effect of exercise on protein turnover in isolated
     soleus and extensor digiitorium longus muscles. Experientia *41:* 1399–1440
     (1985).
54   Kasperek, G.J.; Snider, R.D.: Effect of exercise intensity and starvation on activa-
     tion of branched-chain keto acid dehydrogenase by exercise. Am. J. Physiol. *252:*
     E33–E37 (1987).
55   Keul, J.; Doll, E.; Keppler, D.: Energy metabolism in human muscle, pp. 174–182
     (Karger, Basel 1972).
56   Laurent, G.J.; Millward, D.J.: Protein turnover during skeletal muscle hypertrophy.
     Fed. Proc. *39:* 42–47 (1980).
57   Lefebvre, P.; Luychx, A.; Robaye, B.: Pattern of twenty-four plasma amino acids in
     rats before and after muscular exercise. Archs int. Physiol. Biochim. *80:* 935–940
     (1972).
58   Lemon, P.W.R.; Mullin, J.P.: Effect of initial muscle glycogen levels on protein
     catabolism during exercise. J. appl. Physiol. *48:* 624–629 (1980).
59   Lemon, P.W.R.; Nagle, F.J.: Effects of exercise on protein and amino acid metabo-
     lism. Med. Sci. Sports *13:* 141–149 (1981).
60   Lemon, P.W.R.; Nagle, F.J.; Mullin, J.P.; Benevenga, N.J.: In vitro leucine oxida-
     tion at rest and during two intensities of exercise. J. appl. Physiol. *53:* 947–954
     (1982).
61   Lemon, P.W.R.; Dolny, D.G.; Yarasheski, K.E.: Effect of intensity on protein utili-
     zation during prolonged exercise (Abstract). Med. Sci. Sports Ex. *16:* 151 (1984).
62   Liebig, J. von: Animal chemistry or organic chemistry in its application to physiol-
     ogy and pathology (Taylor & Walton, London 1842).
63   Loughna, P.; Goldspink, G.; Goldspink, D.F.: Effect of inactivity and passive stretch
     on protein turnover in phasic and postural rat muscles. J. appl. Physiol. *61:* 173–179
     (1986).
64   Lowenstein, J.M.: Ammonia production in muscle and other tissues: the purine
     nucleotide cycle. Physiol. Rev. *52:* 382–414 (1972).
65   Lundholm, K.; Edström, S.; Ekman, L.; Karlberg, I.; Walker, P.; Scherstein, T.:
     Protein degradation in human skeletal muscle tissue: the effect of insulin, leucine,
     amino acids and ions. Clin. Sci. *60:* 319–326 (1981).

66   McGilvery, R.W.; Goldstein, G.W.: Biochemistry. A functional approach, pp. 608–611 (Saunders, London 1983).

67   Meredith, C.; Bier, D.M.; Meguid, M.M.; Matthews, D.E.; Wen, Z.; Young, V.R.: Whole body amino acid turnover with 13C tracers: a new approach for estimation of human amino acid requirements; in Wesdorp, Soetens, Clinical nutrition 1981, pp. 42–59 (Churchill Livingstone, Edinburgh 1982).

68   Meyer, R.A.; Dudley, G.A.; Terjung, R.L.: Ammonia and IMP in different skeletal muscle fibers after exercise in rats. J. appl. Physiol. *49:* 1037–1041 (1980).

69   Meyer, R.A.; Terjung, R.L.: Differences in ammonia and adenylate metabolism in contracting fast and slow muscle. Am. J. Physiol. *237:* C111–C118 (1979).

70   Meyer, R.A.; Terjung, R.L.: AMP deamination and IMP reamination in working skeletal muscle. Am. J. Physiol. *239:* C32–C38 (1980).

71   Millward, D.J.; Bates, P.C.; Grimble, G.K.; Brown, J.G.; Nathan, M.; Rennie, M.J.: Quantitative importance of non-skeletal muscle sources of N-methylhistidine in urine. Biochem. J. *190:* 225–228 (1980).

72   Millward, D.J.; Davies, C.T.M.; Halliday, D.; Wolman, S.L.; Matthews, D.; Rennie, M.: Effect of exercise on protein metabolism in humans as explored with stable isotopes. Fed. Proc. *41:* 2686–2691 (1982).

73   Molé, P.A.; Baldwin, K.M.; Terjung, R.L.; Holloszy, J.O.: Enzymatic pathways of pyruvate metabolism in skeletal muscle: adaptations to exercise. Am. J. Physiol. *224:* 50–54 (1973).

74   Narahara, H.; Holloszy, J.O.: The actions of insulin, trypsin and electrical stimulation on amino acid transport in muscle. J. biol. Chem. *249:* 5435–5443 (1974).

75   Newsholme, E.A.; Leech, A.R.: Biochemistry for the medical sciences (Wiley, Chichester 1983).

76   Odessey, R.; Khairallah, E.A.; Goldberg, A.: Origin and possibe significance of alanine production by skeletal muscle. J. biol. Chem. *249:* 7623–7629 (1974).

77   Parkhouse, W.S.; McKenzie, D.C.; Hochachka, P.W.; Ovallee, W.K.: Buffering capacity of deproteinized human vastus lateralis muscle. J. appl. Physiol. *58:* 14–17 (1985).

78   Pettenkofer, M.; Voit, C.: Untersuchungen über den Stoffverbrauch des normalen Menschen. Z. Biol. *2:* 459 (1866).

79   Poortmans, J.: Effects of long-lasting physical exercise and training on protein metabolism; in Howald, Poortmans, Metabolic adaptation to prolonged physical exercise, pp. 212–228 (Birkhäuser, Basel 1975).

80   Poortmans, J.: Protein turnover during exercise; in Landry, Orban, Regulatory mechanisms in metabolism during exercise, pp. 159–184 (Symposia Specialists, Miami 1978).

81   Poortmans, J.R.: Protein turnover and amino acid oxidation during and after exercise. Med. Sport Sci. *17:* 130–147 (1984).

82   Poortmans, J.R.: Use and usefulness of amino acids and related substances during physical exercise; in Benzi, Packer, Siliprandi, Biochemical aspects of physical exercise, pp. 285–294 (Elsevier, Amsterdam 1986).

83   Poortmans, J.: In Di Prampero, Poortmans, Protein metabolism. Effects of exercise and training, pp. 66–76 (Karger, Basel 1981).

84   Poortmans, J.; Delisse, L.: The effect of graduated exercise on venous pyruvate and alanine in humans. J. Sports Med. phys. Fitness *17:* 123–130 (1977).

85 Poortmans, J.; Siest, G.; Galteau, M.M.; Houot, O.: Distribution of plasma amino acids in humans during submaximal prolonged exercise. Int. Z. angew. Physiol. Arbeitsphysiol. *32:* 143–147 (1974).

86 Radha, E.; Bessman, S.P.: Effect of exercise on protein degradation 3-methylhistidine and creatinine excretion. Biochem. Med. *29:* 96–100 (1983).

87 Refsum, H.E.; Gjessing, L.R.; Strömme, S.B.: Changes in plasma amino acid distribution and urine amino acids excretion during prolonged heavy exercise. Scand. J. clin. Lab. Invest. *39:* 407–413 (1979).

88 Refsum, H.E.; Strömme, S.B.: Urea and creatinine production and excretion in urine during and after prolonged heavy exery exercise. Scand. J. clin. Lab. Invest. *35:* 775–780 (1974).

89 Rennie, M.J.; Edwards, R.H.T.; Davies, M.; Krywawych, S.; Halliday, D.; Waterlow, J.C.; Millward, D.J.: Protein and amino acid turnover during and after exercise. Biochem. Soc. Trans. *8:* 499–501 (1980).

90 Rennie, M.J.; Edwards, R.H.T.; Krywawych, S.; Davies, C.T.M.; Hallyday, D.; Waterlow, J.C.; Millward, D.J.: Effect of exercise on protein turnover in man. Clin. Sci. *61:* 627–639 (1981).

91 Rennie, M.J.; Millward, D.J.: 3-Methylhistidine excretion and 3-methylhistidine/creatinine ratio are poor indicators of skeletal muscle protein breakdown. Clin. Sci. *65:* 217–225 (1983).

92 Rogers, P.A.; Jones, G.H.; Faulkner, J.A.: Protein synthesis in skeletal muscle following acute exhaustive exercise. Muscle Nerve *2:* 250–256 (1979).

93 Ruderman, N.B.; Lund, P.: Amino acid metabolism in skeletal muscle. Regulation of glutamine and alanine release in the perfused rat hindquarter. Israel J. med. Scis *8:* 295–302 (1972).

94 Sahlin, K.; Henriksson, J.: Buffer capacity and lactate accumulation in skeletal muscle of trained and untrained men. Acta physiol. scand. *122:* 331–339 (1984).

95 Siesjo, B.K.; Messeter, K.: In Siesjo, Sörensen, Ion homeostasis of the brain (Munksgaard, Copenhagen 1971).

96 Swartman, J.R.; Taylor, P.B.; Cook, B.: Effect of exercise on amino acid incorporation into myocardial contractile proteins. Pflügers Arch. *391:* 319–323 (1981).

97 Tischler, M.E.; Fagan, J.M.: Relationship of the reduction-oxidation state to protein degradation in skeletal and atrial muscle. Archs Biochem. Biophys. *217:* 191–201 (1982).

98 Vazquez, J.A.; Paul, H.S.; Adibi, S.A.: Relation between plasma and tissue parameters of leucine metabolism in fed and starved rats. Am. J. Physiol. *250:* E615–E621 (1986).

99 Viru, A.: Mobilisation of structural proteins during exercise. Sports Med. *4:* 95–128 (1987).

100 Wahren, J.; Felig, P.; Hendler, R.; Ahlborg, G.: Glucose and amino acid metabolism during recovery after exercise. J. appl. Physiol. *34:* 838–845 (1973).

101 Waterlow, J.C.; Garlick, P.J.; Millward, D.J.: Protein turnover in mammalian tissues and in the whole body (Elsevier, Amsterdam 1978).

102 Waterlow, J.C.; Jackson, A.A.: Nutrition and protein turnover in man. Br. med. Bull. *37:* 5–10 (1981).

103 Wenger, H.A.; Wilkinson, J.G.; Dallaire, J.; Nihei, T.: Uptake of ^{3}H-leucine into different fractions of rat skeletal muscle following acute endurance and sprint endurance. Eur. J. appl. Physiol. *47:* 83–92 (1981).

104 White, T.P.; Brooks, G.A.: (U[14]C)glucose, -alanine, and -leucine oxidation in rats at rest and two intensities of running. Am. J. Physiol. *240:* E155–E165 (1981).

105 Wilkerson, J.E.; Batterton, D.L.; Horvath, S.M.: Exercise induced changes in blood ammonia levels in humans. Eur. J. appl. Physiol. *37:* 255–263 (1977).

106 Williams, J.N.; Schurr, P.E.; Elvehjem, C.A.: Influence of chilling and exercise on free amino acid concentrations in rat tissues. J. biol. Chem. *182:* 55–59 (1950).

107 Wolfe, R.R.; Goodenough, R.D.; Wolfe, M.H.; Royle, G.T.; Nadel, E.R.: Isotopic analysis of leucine and urea metabolism in exercising humans. J. appl. Physiol. *52:* 458–466 (1982).

108 Wolfe, R.R.; Wolfe, M.H.; Nadel, E.R.; Shaw, J.H.F.: Isotopic determination of amino acid-urea interactions in exercise in humans. J. appl. Physiol. *56:* 221–229 (1984).

109 Young, V.R.: Skeletal muscle and whole-body protein metabolism in relation to exercise; in Poortmans, Niset, Biochemistry of exercise, sect. IV A, pp. 59–74 (University Park Press, Baltimore 1981).

110 Young, V.R.; Munro, H.N.: 3-Methylhistidine and muscle protein turnover. An overview. Fed. Proc. *37:* 2291–2300 (1978).

Dr. Jacques R. Poortmans, Chimie Physiologique, ISEPK-CP 168,
Université Libre de Bruxelles, avenue Paul-Héger 28, B–1050 Bruxelles (Belgium)

Poortmans JR (ed): Principles of Exercise Biochemistry.
Med Sport Sci. Basel, Karger, 1988, vol 27, pp 194–211.

Application of Knowledge of Metabolic Integration to the Problem of Metabolic Limitations in Sprints, Middle Distance and Marathon Running

E.A. Newsholme

Department of Biochemistry, University of Oxford, UK

Introduction

Knowledge of the factors involved in the provision of energy for muscles and those that might limit performance has accrued largely through the work of physiologists and biochemists who are primarily interested in the chemical, biochemical and biophysical aspects of energy provision. Another field of biochemistry, which has only recently been applied to exercise, is that of regulation and integration of metabolism and metabolic pathways. Application of knowledge from this field to the question of the control of fuel supply for exercising muscle leads to some new insights into metabolic limitations in sprinting, middle distance and marathon running.

Importance of the Maintenance of the ATP/ADP Concentration Ratio in Muscle

It is known that ATP does not function as a store of chemical energy in the cell. Its concentration in muscle is only 5–7 μmol/g fresh muscle [4] which would be depleted in less than a second during intense muscular activity unless it was resynthesised at a rate equal to that of utilisation. In combination with ADP, ATP functions as an energy transfer system in the cell: the generation of ATP from ADP during the oxidation of fuels (e.g.

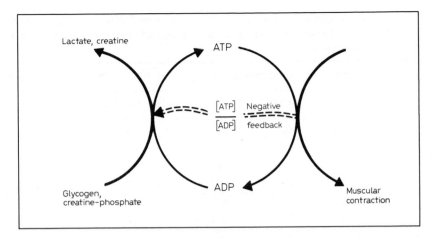

Fig. 1. ATP links energy-producing processes to those that require energy. As both processes take place at the same rate, the concentration of ATP remains fairly constant. Any small change in the ratio (ATP/ADP) acts as a signal to change the rate of ATP formation to match that of ATP utilisation.

glucose) conserves chemical energy, which is utilised in a number of processes [4] (e.g. muscular contraction). The ATP-ADP system couples the energy-releasing processes of the cell with the contractile process in such a way that the latter process is totally dependent upon the former. Thus, when contractile activity is increased, the rate of fuel utilisation must also be increased. Furthermore, to avoid large transient changes in the ATP/ADP concentration ratio, the rate of fuel oxidation must be regulated *rapidly* and *precisely* according to the rate of ATP utilisation by the contractile process (fig. 1). Indeed, the ratio of energy production to energy utilisation is known to remain constant under these conditions. For example, sprinting in man increases the rate of ATP turnover about 1,000-fold but it results only in small changes in the ATP concentration (table I) and in the ATP/ADP concentration ratio [2, 4].

The reason for this remarkable constancy of the ATP/ADP ratio may be to maintain what the author terms the 'kinetic efficiency' of the energy-producing and energy-utilising processes in the muscle. Since ATP is chemically very similar to ADP, the catalytic site of enzymes that react with these nucleotides cannot totally distinguish between them. This lack of complete discrimination manifests itself as competitive inhibition of the enzyme activity by each nucleotide in relation to the other. Thus, enzymes

Table I. The concentration of muscle ATP and phosphocreatine, blood lactate and the blood pH in sprinters during the sprint: data are presented as means from 3 runners taken from Hirvonen et al. [2]

Distance run, m	Time	Concentration, mM			Capillary blood pH
		ATP	phosphocreatine	blood lactate	
40	before	5.4	10.3	1.5	7.44
	after	3.5	3.8	4.5	7.36
60	before	5.5	10.8	1.5	7.42
	after	3.2	4.1	5.9	7.31
80	before	5.4	10.3	1.5	7.41
	after	3.3	2.5	6.8	7.28
100	before	5.2	9.1	1.6	7.42
	after	3.7	2.6	8.3	7.24

that utilise ADP as substrate are inhibited by ATP and enzymes that utilise ATP as substrate are inhibited by ADP (table II). Consequently, if the intracellular ATP/ADP concentration ratio is increased, the activity of enzymes catalysing the conversion of ADP into ATP would be inhibited, whereas if the ratio decreased, the activity of enzymes catalysing ATP utilisation would be inhibited. It is likely that small changes in this concentration ratio in the cell would cause only slight inhibition, but any decrease in the activity of enzymes catalysing *important* regulatory reactions in the cell could reduce the rate of ATP formation and/or the performance of mechanical work. In this way, the kinetic efficiency of energy transfer in the muscle would be reduced. Although a reduction in this kinetic efficiency may have little obvious effect on the everyday life of a 'normal' human subject enjoying the benefits of civilisation, it could seriously interfere with the performance of an athlete. The challenge of competition demands that the formation, transfer and utilisation of biological energy in the muscle occurs with maximum efficiency. In other words, changes in the concentration of intracellular adenine nucleotides, and also all other metabolic intermediates, should be minimal during changes in the rate of energy utilisation in muscle.

The changes in rate of metabolism, the increases in oxygen consumption and the long hard hours of training are all designed to maintain the

Table II. Inhibition of ATP-producing and ATP-utilizing enzyme activities by adenine nucleotides [data from ref. 4]

Enzyme	Substrate	Inhibitor	Type of inhibitor
Hexokinase	ATP	ADP	competitive
3-Phosphoglycerate kinase	ADP	ATP	competitive
Pyruvate kinase	ADP	ATP	competitive
Creatine phosphokinase	ATP	ADP	competitive
Phosphoenol pyruvate carboxylase	GTP	GDP	mixed type
Fatty acyl-CoA synthetase	ATP	ADP	competitive
Adenylate kinase	ATP	ADP	competitive
Actomyosin ATPase	ATP	ADP	
Adenine nucleotide translocase	ADP	ATP	

The inward transport of ADP by heart muscle mitochondrial translocase is not specific for ADP so that an increase in the ATP concentration in the cytoplasm could reduce the activity of the translocase.

ATP/ADP concentration ratio as constant as possible despite dramatic changes in the rate of ATP utilisation. Fatigue is the effect or indeed perhaps the anticipation that this ratio will decrease sufficiently to damage the muscle. However, fatigue is achieved in different ways and these will be presented below in relation to control in different types of running activity.

Metabolic Integration and Limitation in Sprinting

Maximum sprinting utilises about 3 μmol of ATP per gram of active muscle each second [4]. It demands the utilisation of either 1 μmol (or 0.4 mg) of glucose from glycogen or 3 μmol of creatine phosphate each second.

The concentration of glucose as glycogen in human muscle is approximately 80 μmol per gram and is probably somewhat higher than this in the muscle of elite sprinters. This could provide, in theory, 240 μmol of ATP, which is sufficient energy for 80 s of sprinting [4]. But maximum power output (i.e. that achieved in the 100-metre sprint) cannot be maintained for 80 s. In other words, the power output is reduced as the duration of the race increases above 100 m. The explanations for the fatigue in these events are given in other chapters.

One further factor that the author considers can limit performance in the sprint is the ability to increase rapidly the rate of glycolysis from the low values at rest to the high values during activity. One means of doing this is via substrate (futile) cycles.

Metabolic Control of a Key Reaction in Glycolysis

The rate of glycolysis (glycogen conversion to lactate) in resting muscle is about 0.05 µmol/min. It increases to a maximum of 50–60 µmol/min/g muscle during sprinting. How is this enormous increase in rate achieved? Feedback control through a mechanism that provides high sensitivity in metabolic control may be part of the answer.

A key reaction in glycolysis, the phosphorylation of fructose 6-phosphate, is catalysed by the enzyme phosphofructokinase (PFK), which plays an important role in the control of the rate of glycolysis. Contraction results in an increase in the rate of conversion of ATP to ADP which changes their concentrations such that the ATP/ADP concentration ratio decreases. It is the change in this ratio that is largely responsible for providing feedback control between the rate of ATP utilisation and the activity of PFK and hence the rate of glycolysis (fig. 2). However, maintenance of the ATP/ADP concentration ratio near to normal is required for kinetic efficiency so that *large* changes in the concentrations of these energy nucleotides cannot occur. Consequently, a small decrease in the ATP/ADP concentration ratio must produce the enormous increase in the catalytic activity of PFK during sprinting. This can be achieved *only* if the regulatory system is *very* sensitive, so that small changes in the concentrations of these regulators produce a large response in the rate of the enzyme-catalysed reaction. This high sensitivity is achieved by the presence of a further enzyme, fructose bisphosphatase (FBPase) which catalyses the reverse reaction to that of PFK, i.e. dephosphorylation of fructose bisphosphate: the simultaneous activities of the two enzymes produce a cyclic flux from fructose 6-phosphate to fructose bisphosphate which is known as a substrate cycle. Such cycles provide high sensitivity only if the rate of cycling compared to the net flux through the pathway is high [3]. The response of the cycling and glycolytic systems to the race is seen as follows.

When the sprinter is resting in the changing room prior to the race, the activities of both PFK and FBPase are thought to be very low. When the sprinter is on his blocks waiting for the gun, the stress hormones, adrenaline and noradrenaline, cause a stimulation of the catalytic activities of both enzymes, so that the rate of cycling is high. But, since the muscles are

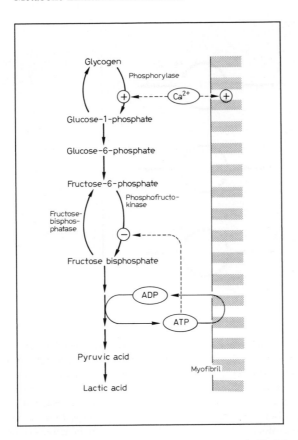

Fig. 2. In this pathway of glycolysis only about half of the intermediates are named. Each is formed from the previous one by the action of a different enzyme. Control steps are shown with a broken line. Ca^{2+} control both the myofibrils (i.e. cross-bridge cycling – the myofibrillar ATPase) and the activity of phosphorylase – the flux-generating step for anaerobic glycolysis in muscle. The ATP/ADP concentration ratio controls the activity of phosphofructokinase.

not mechanically active, the energy demand is small and hence the glycolytic flux is very low. In other words, the flux through the PFK reaction is almost 'balanced' by the flux through the reverse reaction (FBPase) so that the *net* flux is low [4]. This condition provides the high sensitivity; only a small change in the ATP/ADP concentration ratio is needed to produce an enormous increase in the *net* flux. This change occurs, of course, when the gun is fired and the race begins (fig. 3).

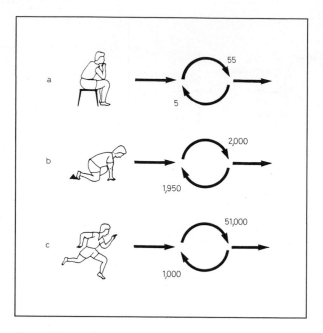

Fig. 3. Changes in the rate of cycling and flux through glycolysis which might occur (a) before a 100-metre sprint; (b) at the start of the 100-metre sprint, and (c) during a 100-metre sprint. The upper arrow represents the enzyme phosphofructokinase and the lower arrow the enzyme fructose bisphosphatase. The flux is the difference between the rates of the forward and reverse reactions. The stress hormones are considered to be responsible for activation of both enzymes in situation (b) – and changes in the extent of ATP feedback control are considered to be responsible for the change in activity of phosphofructokinase from state (b) to state (c) (fig. 2).

Elite sprinters probably have a high capacity for cycling and it is predicted that the changes in stress hormones increase cycling rates close to the theoretical maximum, when required: i.e. when the sprinter is on the blocks waiting for the gun during a competitive race. The greater the cycling rate, the greater is the sensitivity in control and the smaller the change in ATP/ADP concentration ratio required to produce near maximum rates of glycolysis. This ensures that high glycolytic rates can be maintained during the race. Under this condition, only the decrease in the phosphocreatine concentration and/or the accumulation of hydrogen ions, and not changes in the ATP/ADP concentration ratio, will interfere in the contraction process.

Metabolic Integration and Limitations in Middle Distance Running

Runners of distances below about 3,000 m probably make use of both aerobic and anaerobic processes; that is, glucose or glycogen will be completely oxidised via the Krebs cycle but, in addition, some glycogen will be degraded to lactate. Some lactate and proton production can occur from anaerobic glycolysis without the risk of fatigue since the better blood supply to the muscle of middle distance runners in comparison to sprinters will allow continuous removal of lactate plus protons into the blood. In the blood, the protons can be effectively buffered and eventually removed as carbon dioxide in the lungs and the lactate removed by oxidation in other muscles such as the heart or reconverted to glucose via gluconeogenesis in the liver. This will allow the sprint type of fibres to contribute to the power output of the middle distance runner and, since they have a greater power output than the endurance fibres, it may be advantageous for middle distance runners to possess muscles with both types of fibres.

Marathon Running

The distance of the marathon race is 42.2 km (26 miles and 385 yards) which is completed by elite runners in about 130 min, at an average energy expenditure of almost 12,000 kJ. Even non-elite runners are able to complete the run in about 150–170 min and the total energy expenditure is similar. Although there are respiratory and cardiovascular requirements for the athlete to be able to maintain such a high power output for a long period of time, recent work has indicated that fuel supply for the muscles and its control are major problems for both elite and non-elite runners.

Fuels Available to the Muscles

The ATP production for the marathon runner is achieved by the oxidation of the two major fuels, carbohydrate and fat. The amount of carbohydrate and fat stored in the body, where they are stored, how they are mobilised, and how the rates of utilisation of the two fuels are integrated are important in understanding the problems of fuel supply facing the marathon runner.

The largest fuel reserve in the body is fat (triacylglycerol) which, in theory, could ensure fuel supply for about 119 h of marathon running [4]. Triacylglycerol is stored in adipose tissue which is distributed diffusely throughout the body; for example, under the skin, around the major organs and in the peritoneal cavity. The triacylglycerol in the fat cell is hydrolysed to fatty acids which enter the blood and are transported to muscle (and other tissues) in combination with albumin. The reaction is as follows:

fatty acid + albumin $\rightleftarrows$ fatty acid – albumin.

This is necessary since the solubility of fatty acids in an aqueous medium is very low (in principle, it is similar to the transport of oxygen in combination with haemoglobin). This transport mechanism is very effective in that a large quantity of fat can be transported, but it probably sets an upper limit on the rate at which fatty acid oxidation can occur in muscle. The rate of diffusion of fatty acid from the albumin complex into the muscle cell is governed by the *free* (i.e. that not bound to albumin) concentration of fatty acid, which is very low. Hence, the rate of oxidation of fatty acid is also low. Evidence suggests that this low rate of diffusion restricts fatty acid oxidation by muscle to such an extent that it can only provide a proportion, *about 50%,* of the maximum demand for energy by muscle [4].

(a) If the body's store of carbohydrate is depleted prior to a run (e.g. by eating a low-carbohydrate diet) endurance is decreased: that is, despite the large stores of fat, exhaustion occurs very early in an endurance event after the runner has been on a low-carbohydrate diet.

(b) If the body's store of carbohydrate is increased prior to a run (e.g. by eating a high-carbohydrate diet) endurance is increased. Most marathon runners attempt various dietary tricks to increase the glycogen level in the muscle prior to a run.

(c) In ultradistance running (that is, distances greater than the marathon – 50 miles, 100 miles, etc.) the power output gradually declines to about 50% of the maximum as fat becomes the dominant fuel in the later stages of the run.

(d) The maximum ability to take up oxygen during running is about half of what is expected in patients suffering from an inability to oxidise glucose and who are, therefore, dependent solely on fatty acid oxidation.

The general conclusion is, therefore, that fatty acids *are* used during sustained running but that the maximum rate at which they can be used is limited to *about* 50% of the maximum rate of energy utilisation required to support maximum running speed. It is possible that this proportion is higher in elite marathon runners and in ultramarathon runners. The difference is made up by oxidation of carbohydrate, either in the form of muscle glycogen or glucose from the blood stream.

Controlling the Correct Mixture of Fuels

When blood glucose (or muscle glycogen) and fatty acids are both available to the muscle, the fatty acids are oxidised in preference to the glucose. The control system known as the glucose/fatty acid cycle [4] ensures that as much fatty acid oxidation as possible occurs, up to *about* 50% of the maximum oxygen uptake, but that the remainder of the energy is generated by the oxidation of blood glucose or muscle glycogen. It is a mechanism that always allows a mixture of fuels, carbohydrate and fat, to be burnt *provided* that both are available. The advantage of this is that it 'spares' carbohydrate so that the limited glycogen will last longer.

It is undoubtedly this mechanism which is at least partly responsible for the maintenance of the blood level at about 4–5 m*M* despite a massive demand for fuel by the exercising muscle. Hence, any decrease in the rate of mobilisation of fatty acids would increase the rate of glucose utilisation and could, therefore, result in hypoglycaemia.

Exhaustion occurs in the marathon runner when the glycogen store in the muscle is depleted. This means that ATP can no longer be regenerated from the oxidation of carbohydrate, it can only be obtained from the oxidation of fatty acid. But the latter can only provide about half the energy required; hence, the pace must be decreased by about 50%. This means that, for an elite runner with a normal pace of 11–12 miles per hour, the pace will fall to 5–6 miles per hour or just above fast walking. This, for the elite runner, is exhaustion. What happens if the athlete continues to run at the high pace despite the exhaustion; provided the athlete can withstand the pain of exhaustion, blood glucose will be used. But with little or no liver glycogen at this time, the imminent danger is hypoglycaemia; the symptoms of hypoglycaemia include nausea, weakness, sweating, palpitations, a feeling of detachment from the environment, visual disturbances and unsteady gait. Hypoglycaemia can occur both during and after the race. The effects of hypoglycaemia in marathon runners were dramatically recorded in 1984 at the Los Angeles Olympic Games; when the Swiss marathon runner Gabriela Anderson-Schiess entered the Coliseum she lurched from side to side and appeared to be in a state of total fatigue. She was suffering from hypoglycaemia. After completion of the race, she was given intravenous glucose and rapidly recovered.

The long training runs (20 plus miles) are so necessary for the marathon runner since this trains the metabolic mechanism for the precise mobilisation of fatty acids from adipose tissue and the metabolic mechanism for regulating the rate of glucose and glycogen utilisation in relation to the rate of fatty acid oxidation. Failure of either of these mechanisms could result in higher rates of utilisation of glycogen and hence an earlier onset of fatigue. The integration between carbohydrate and fat metabolism also explains one further fact which is well known to marathon runners and coaches: a period of starvation for at least 4 h (some runners prefer to starve for 12 h) is advantageous – only water should be taken. Even fluid containing sugar or carbohydrate should be avoided. The reason for this is that ingestion of any carbohydrate increases the rate of secretion of insulin which raises the plasma level of this hormone and this prevents the mobilisation of fatty acid from adipose tissue. This means that more glycogen has to be used in the early part of the marathon – so increasing the risk that glycogen stores will be depleted well before the end of the marathon.

Control of Fatty Acid Mobilisation by Adipose Tissue

Fatty acid oxidation provides a considerable proportion of the energy required for prolonged severe exercise (e.g. marathon running). Although the blood fatty acid level is one important factor controlling the rate of fatty acid oxidation, the energy demand of the muscle will undoubtedly play some part in the regulation of fatty acid uptake and oxidation. However, the biochemical mechanism of this latter control is unknown. Consequently, discussion in this section will centre upon the mechanisms by which the rate of fatty acid mobilisation is regulated precisely according to the rate of utilisation, so that a steady state condition is achieved.

Hormonal Control of Lipolysis in Adipose Tissue
There are a large number of hormones that can increase the rate of lipolysis, but only insulin (and certain prostaglandins) inhibit lipolysis. During exercise, increases in the blood levels of the lipolytic hormones, adrenaline, glucagon and possibly noradrenaline, together with a decrease in the level of insulin may be responsible, at least in part, for the increase in lipolytic rate [4]. However, the author considers that changes in levels of hormones (or nervous stimulation) cannot provide sufficiently precise control of the rate of fatty acid mobilisation to satisfy the energy requirements of the muscle: it seems highly unlikely that endocrine glands or even higher nervous centres can accurately monitor the work being done by muscular tissue at any instant of time. Nonetheless, the establishment of a steady

state rate for the pathway of fatty acid utilisation would appear to be fundamentally important for the metabolic well-being of the individual. Thus, if the rate of fatty acid mobilisation exceeded that of utilisation, the blood fatty acid level would increase continuously, such that it might reach levels that could cause damage to various tissues. High concentrations of fatty acids damage cell and mitochondrial membranes, uncouple oxidative phosphorylation, increase the rate of platelet aggregation, may result in fibrillation of the heart and could be responsible for hypertriglyceridaemia and eventual atherosclerosis [4]. On the other hand, if the rate of fatty acid mobilisation was less than that of utilisation, the fatty acid level in the blood would decrease which would result in glucose utilisation by the muscle and a fall in the blood glucose level and possibly result in hypoglycaemia (see above). This could cause fatigue, which would be overcome only by a reduction in the energy expenditure by the musculature. Thus, hormonal control of lipolysis must be supplemented by a mechanism that provides a feedback link between the rate of energy utilisation by the muscle and the process of lipolysis. Two such mechanisms, which are described below, may exist.

Ketone Body Control of Fatty Acid Mobilisation

Ketone bodies (acetoacetate and 3-hydroxybutyrate) are produced in the liver by the partial oxidation of fatty acids to acetyl-CoA and conversion of the latter to acetoacetate via the hydroxymethyl glutaryl-CoA (HMG-CoA) cycle [4]. The blood level of fatty acids is one factor in the control of the rate of ketogenesis in the liver, so that if the concentration of fatty acids increases, the rate of ketogenesis increases and the concentration of ketone bodies in the blood is raised. It has been shown that high levels of ketone bodies can increase the sensitivity of adipose tissue for insulin: hence the antilipolytic effect of this hormone is increased so that the rate of fatty acid mobilisation is decreased. It is suggested that these effects provide a sensitive feedback control to relay information concerning the blood level of fatty acids to the lipolytic system. If the rate of fatty acid mobilisation exceeds that of utilisation, the blood level of fatty acids will increase which will stimulate the rate of ketogenesis and increase the blood level of ketone bodies (fig. 4). This process is probably most important when the exercise is finished. The phenomenon of post-exercise ketosis may play such a role. The author suggests that, during exercise, another mechanism may play a more important role in control of fatty acid mobilisation: this is the triglyceride-fatty acid cycle in adipose tissue.

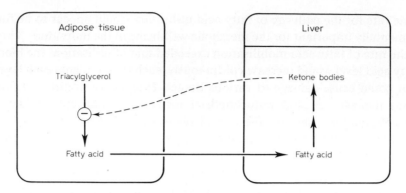

Fig. 4. Feedback inhibition by ketone bodies on lipolysis in adipose tissue. This is important, particularly when exercise has ceased, to prevent excessive increase in the plasma FFA level.

Feedback Control of Fatty Acid Mobilisation via the Triglyceride/Fatty Acid Cycle

In adipose tissue, the process of lipolysis occurs simultaneously with that of esterification, so that triglyceride is broken down to fatty acids which are reactivated and re-esterified to form triglyceride (i.e. a substrate cycle) [3]. One advantage of a substrate cycle is the improvement in sensitivity of metabolic control mechanisms, and this may be particularly important for the effects of hormones and ketone bodies on lipolysis. Thus, small changes in the activity of the triglyceride lipase could produce large changes in the rate of fatty acid mobilisation. However, the properties of the esterification and lipolytic processes permit the formulation of a theory of control of lipolysis by changes in the blood level of fatty acids. If the rate of fatty acid utilisation by muscle increased (due to increased work), this would decrease the concentration of fatty acids in the blood and hence that in adipose tissue. Since the concentration of fatty acid in adipose tissue is not saturating for the esterification process, a decrease in concentration would reduce the rate of esterification. Hence, the rate of fatty acid mobilisation would increase. Furthermore, since fatty acids inhibit the activity of triglyceride lipase, a decrease in the fatty acid concentration would increase the rate of lipolysis. Consequently, an increased rate of utilisation of fatty acids in muscle, causing a decrease in fatty acid concentration in adipose tissue, will lead automatically to an increased rate of mobilisation

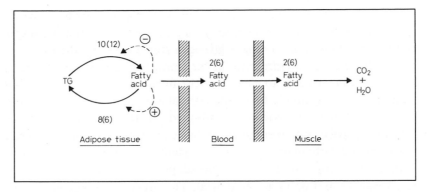

Fig. 5. Feedback control of fatty acid mobilisation from adipose tissue via the triglyceride-fatty acid cycle. The numbers represent the (hypothetical) rates of the various processes. The numbers in parentheses indicate the changes in rates produced by an increased rate of fatty acid utilisation by muscle. Thus, a decrease in the plasma fatty acid level is expected to decrease the rate of esterification and increase that of lipolysis.

of fatty acids from adipose tissue. Of course, the opposite changes would result from a decreased rate of utilisation of fatty acids in muscle (fig. 5).

Athletic training may stimulate the capacity of such cycles and increase the cycling/flux ratio so that an exquisitely sensitive control mechanism is produced.

Manipulating Metabolism

Knowledge of the oxidative processes producing energy for the marathon runner and of the metabolic limitations enable a number of suggestions to be put forward to improve performance, or at least not to impair it!

(i) To ensure that the glycogen stores are as high as possible prior to the race, the regime of 'glycogen unloading, glycogen loading' has been recommended. However, for a runner training more than about 50 miles (80 km) per week, this may be replaced by the 'glycogen loading' regime. This simply means that a large quantity of carbohydrate should be eaten in the 2–4 days before the race, depending upon the individual. The precise time will need to be ascertained by trial and error – loose bowels and excessive urination usually mean the peak glycogen level has passed. In what form should this extra carbohydrate be taken? The normal response of the body to food is to replete glycogen stores first and then to convert any additional carbohydrate to fat for storage in the adipose tissue. The marathon runner needs to encourage the former but discourage the latter. This can be done simply by eating normal amounts of food but dramatically increasing the proportion of carbohydrate; the best pattern is three or four meals spaced out over the day with a couple of snacks in between. The nature of the carbohydrate is also important. It should be complex carbohydrate, not simple sugars, so that bread, potatoes, cereals, rice, pearl barley, spaghetti and pasta are beneficial. The intake of food containing the simple sugars, chocolate bars, sweetened drinks, honey, jam, should not be excessive and only accompany the complex carbohydrate. The rate of digestion of complex carbohydrates is slow and hence the absorption of sugar into the body takes place over a long time and this encourages the storage of glycogen instead of fat. The rate of digestion depends not only on the state of the carbohydrate but also on the presence of fat and protein. These will slow down passage through the small intestine and favour the slow, gradual absorption of glucose. Hence meat, cheese, fish, egg should be eaten at each meal but less than usual, since carbohydrate must replace much of the protein.

As indicated earlier, glycogen is stored in association with water so that during the 'loading' phase it may be necessary to increase the intake of liquid to avoid dehydration. A simple means of testing this is to observe the colour of the urine – it should be clear not amber.

The large quantity of glycogen plus water in the muscles can give a feeling of heaviness to the legs in the first few miles of the marathon but if the explanation is known it should dispel anxiety. Furthermore, the feeling will soon disappear as the excess glycogen is used and the trapped water released, and the benefits will be felt after 15 miles (24 km).

(ii) If the amount of carbohydrate stored by the body limits the marathon runner, does it not make metabolic sense to eat easily digested carbohydrate (e.g. glucose) just prior to the race? The answer is 'no'. The reason emphasises the integrated nature of metabolic processes in the body. The mobilisation of fatty acids during endurance running is largely due to changes in the blood levels of two important hormones, adrenaline and insulin. The former stimulates fatty acid mobilisation, the latter inhibits it. Adrenaline is released by the adrenal medulla in response to the excitement and stress prior to the start of the race and is further increased during the race. The insulin level, however, decreases as the race proceeds. These changes probably account for the smooth increase in the rate of mobilisation of fatty acids during the run. Eating, especially carbohydrates, causes the release of insulin, in large amounts, which will inhibit fatty acid mobilisation just when it is needed to conserve valuable glycogen stores.

Some runners find it beneficial to eat little or nothing after the evening meal on the day preceding the race, although an intake of fluid should be maintained. Provided that as much rest as possible is taken, this period of starvation will use little glycogen but should encourage fatty acid release from adipose tissue depots. However, such a period of starvation must be from individual choice; many runners find they run better after a good carbohydrate breakfast but this should be taken several hours before the race.

(iii) If it is not recommended to eat carbohydrate before the race, is it advisable to take carbohydrate-containing drinks during the race? The answer is a qualified 'yes'. After about 5 miles (8 km) of running, easily digested carbohydrate (e.g. glucose) can be taken because, at this stage, it does not result in insulin secretion. However, the amount ingested should be limited.

(iv) The control mechanism that operates between fatty acids and glucose will eventually establish a pattern of fuel utilisation during the race in which fatty acid, glucose and glycogen oxidation all contribute to ATP formation. As already explained, it is important to encourage as much fatty acid oxidation as possible and, although this cannot be measured in every individual, by experience from long training runs athletes will get to know the pace they can maintain for long periods. Once this is established in a race it is disadvantageous to vary it unless absolutely necessary (e.g. running uphill) since this will disturb the fuel oxidation pattern and encourage a greater rate of oxidation of the limited carbohydrate stores.

(v) What the runner does at the end of the race is as important for his well-being as his preparation. The race is followed by rest, relaxation and, usually, elation at having completed the distance. This can cause problems because the body is short of carbohydrate and probably short of fluid. The removal of the demand for large amounts of energy by the muscles and the relaxation may decrease the emphasis placed on the control of the blood sugar level by the body so that it can easily fail to protect from the phenomenon of hypoglycaemia. In addition, low blood volume can reduce the flow of blood to the brain as the cardiovascular system readjusts to rest and this will exacerbate any degree of hypoglycaemia. Consequently, the important post-marathon requirements are liquid and carbohydrate. However, as with the pre-run condition, simple carbohydrates can be dangerous; they will now result in insulin secretion, which will lower the blood fatty acid level and force all tissues to use glucose so that severe hypoglycaemia can result. Ideally, the carbohydrate should be complex but in a liquid form (to prevent dehydration). Since the body can now rapidly lose heat, the ideal post-marathon activity is drinking hot thick vegetable soup slowly while wrapped in a warm tracksuit or blanket.

Central Effects of Exercise: Mood, Fatigue and Sleep

Most participants in endurance exercise will claim that the activity provides a feeling of well-being, an improvement in mood and also an overall feeling of tiredness and exhaustion which usually results in a 'good night's sleep'. The improvement of mood is so much a part of the current running boom that the phenomenon of 'runner's high' has become a popular talking point. But the intriguing question is whether the effects of exercise on mood and general tiredness and exhaustion are biochemically related? A possible metabolic link is explored below.

Information in nerves is carried by electrical impulses but when one nerve communicates with another this is usually achieved via specific chemicals. Such chemicals are known as neurotransmitters and there are probably more than 40 such compounds in the brain [4]. Some of the best known include dopamine and noradrenaline, which are synthesised from the amino acid tyrosine, and 5-hydroxytryptamine which is synthesised from the amino acid tryptophan. These amine neurotransmitters are considered to play an important role in the regulation of mood and sleep. Thus, it is considered that a decreased concentration of the monoamines (noradrenaline, dopamine and 5-hydroxytryptamine) in specific areas of the brain results in depression; anti-depressive agents are known to increase the concentration of these amines in the brain and so improve sleep.

Two separate but well-established facts can be brought together to provide a speculative explanation for a relationship between exercise, mood and sleep. First, muscle is the major site of utilisation of three amino acids, valine, leucine and isoleucine, which are known collectively as the branched-chain amino acids. The rate of utilisation of these amino acids by muscle may increase with exercise. Secondly, amino acids are transported into the brain across the blood-brain barrier via specific carriers, and one of these transports the three branched-chain amino acids *plus* tyrosine and tryptophan, which are known as the aromatic amino acids. Since all these five amino acids are normally present in the bloodstream, there is competition to enter the brain via this carrier. However, after exercise, the concentration of branched-chain amino acids in the bloodstream may decrease, due to an increased rate of utilisation of the branched-chain amino acids by muscle. Hence the concentration ratio, aromatic amino acids/branched-chain amino acids, should increase during and after prolonged exercise. This will favour the entry of tyrosine and tryptophan into brain and these increased concentrations could lead to increased concentrations of the neurotransmitters dopamine, noradrenaline and 5-hydroxytryptamine [1].

This could explain the 'central' effects of endurance exercise, that is improving mood and causing a generalised feeling of tiredness or exhaustion. The hypothesis serves to illustrate the importance of the metabolic integration between different organs and emphasises that a full understanding of *all* the effects of exercise can only be obtained by focussing attention on metabolism within *and* between tissues and organs.

References

1   Blomstrand, E.; Celsing, B.; Newsholme, E.A.: Changes in plasma concentrations of aromatic and branched-chain amino acids during sustained exercise in man and their possible role in fatigue. Acta physiol. scand. (in press, 1988).
2   Hirvonen, J.; Rehunen, S.; Rusko, H.; Harkonen, M.: Breakdown of high energy phosphate compounds and lactate accumulation during short supramaximal exercise. Eur. J. appl. Physiol. *56:* 253–259 (1987).
3   Newsholme, E.A.; Crabtree, B.: The role of substrate cycles in metabolic regulation and heat generation. Biochem. Soc. Symp. *41:* 61–110 (1976).
4   Newsholme, E.A.; Leech, A.R.: Biochemistry for the medical sciences (Wiley, Chichester 1983).

E.A. Newsholme, MD, Department of Biochemistry, University of Oxford, South Parks Road, Oxford, OX1 3QU (UK)

Poortmans JR (ed): Principles of Exercise Biochemistry.
Med Sport Sci. Basel, Karger, 1988, vol 27, pp 212–229.

Mechanisms of Muscular Fatigue

R.H. Fitts, J.M. Metzger

Department of Biology, Marquette University, Milwaukee, Wisc., USA

Introduction

The etiology of muscle fatigue is an important question that has interested exercise scientists for over a century. Despite this, a definitive fatigue agent(s) has yet to be identified. However, progress has been made and theories have been developed that, for the most part, explain the experimental findings. The problem is complex as muscle fatigue might result from deleterious alterations in the muscle itself (peripheral fatigue) and/or from changes in the neural input to the muscle. The latter could itself be mediated by changes of central and/or peripheral origin. Furthermore, the cause and degree of muscle fatigue is clearly dependent on the duration, intensity and nature of the exercise, fiber type composition of the muscle, individual level of fitness, as well as numerous environmental factors. For example, fatigue experienced in high intensity, short duration exercise is surely dependent on different factors than those precipitating fatigue in endurance activity. Similarly, fatigue during tasks involving heavily loaded contractions (e.g. weight lifting) will likely differ from that produced during relatively unloaded movement (running, swimming).

The purpose of this review is to discuss muscle fatigue resulting from two general types of activity: short duration, high intensity and endurance exercise. Although no attempt has been made to present a complete review, the important current theories and supportive experimental results will be presented. Details of some of the topics discussed can be found in earlier reviews [16, 21].

Short Duration, High Intensity Exercise

Muscle fatigue is defined, for the purposes of this review, as a loss of force output leading to a reduced performance of a given task. Fatigue during short duration, high intensity exercise could result from an impairment of the central nervous system (CNS), such that the optimal frequency of motor nerve activation is not maintained. Bigland-Ritchie et al. [5] have clearly shown that the frequency of motor nerve firing decreases during continuous contractile activity. The question is whether this change precipitates muscle fatigue or results from neuronal feedback (muscle afferents) in an attempt to maintain an optimal activation frequency as fatigue develops. Recent evidence suggests the latter as the reduced motoneuron firing rate persisted during recovery when the fatigued muscle was kept ischemic [6]. The fatiguing muscle generally shows a prolonged force transient (primarily due to a slower relaxation) in response to a single stimulus, and consequently a lower frequency of activation is required to elicit peak tension (force-frequency curve shifts to the left). It appears that the primary sites of fatigue are located within the muscle, and do not generally involve the CNS, peripheral nerves or the neural-muscular (N-M) junction [16]. The observation that fatigued muscles generate the same tension whether stimulated directly or via the motor nerve argues against N-M junction fatigue [19].

Excitation-Contraction Coupling

Figure 1 shows the major components of a muscle cell involved in excitation-contraction (E-C) coupling. The numbers indicate possible sites within the cell where alteration during short duration, high intensity exercise could induce fatigue.

Surface membrane (fig. 1 No. 1): It has been suggested that fatigue during high frequency stimulation is in part due to an impaired surface membrane action potential transmission which has been hypothesized to result from an accumulation of extracellular potassium and/or a reduction of sodium [15, 16]. With fatigue, the shape of the sarcolemma action potential (AP) changes showing a decreased amplitude and prolonged duration [36]. However, it is unlikely that these alterations directly induce fatigue as force output bears no fixed relation to the size of the AP during either the development of, or recovery from, fatigue [36, 45]. The observation of an uncoupling between membrane electrical properties and force has been demonstrated by a number of investigators [3, 36, 45]. Bezanilla

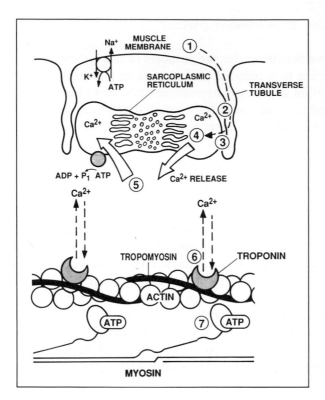

Fig. 1. A diagrammatic representation of the major components of a muscle cell involved in excitation-contraction (E-C) coupling. The numbers indicate possible sites of muscular fatigue during heavy exercise and include: (1) surface membrane; (2) t-tubular charge movement; (3) unknown mechanism coupling t-tubular charge movement with SR Ca^{2+} release; (4) SR Ca^{2+} release; (5) SR Ca^{2+} reuptake; (6) Ca^{2+} binding to troponin; (7) actomyosin hydrolysis of ATP and cross-bridge force development and cycling rate.

et al. [3] fatigued muscles in various concentrations of extracellular sodium. Action potentials declined similarly during stimulation at a time when force was greatly reduced in the low, but not the normal, sodium group [3]. Furthermore, it was the conclusion of Eberstein and Sandow [13] that the relatively small decrease in the amplitude of the AP during repetitive stimulation could not account for the large decline in muscle force (fatigue). The observation that the resting membrane potential was unaltered during fatiguing stimulation [32, 36] further argues against the hypothesis that increases in extracellular potassium mediate fatigue.

The primary sites of muscular fatigue apparently involve processes that occur after depolarization of the sarcolemmal and t-tubular membranes. It is likely that the general process known as excitation-contraction coupling is disrupted. The t-tubular AP produces a charge movement within the t-tubular membrane (fig. 1 No. 2) which subsequently, via an unknown process (fig. 1 No. 3), leads to Ca^{2+} release from the sarcoplasmic reticulum (SR) (fig. 1 No. 4). Recently, it has been suggested that the t-tubular charge movement activates a second messenger system which in turn triggers Ca^{2+} release [53]. The elevated intracellular Ca^{2+} binds to troponin C (fig. 1 No. 6), a regulatory protein, producing a molecular change in the troponin-tropomyosin complex which in turn allows the contractile proteins actin and myosin to bind and generate force [51].

Fatigued muscle frequently shows a prolonged twitch duration and a reduced peak rate of tension development [22, 52]. A prolonged twitch duration reflects a similar prolongation in the time course of the increase in intracellular Ca^{2+} (Ca^{2+} transient). The lengthened Ca^{2+} transient suggests a reduced rate of Ca^{2+} reuptake by the SR and other Ca^{2+}-binding proteins. The decline in the rate of tension development might at least in part be due to a lowered Ca^{2+} release rate. The study of Blinks et al. [8], utilizing the intracellular Ca^{2+} indicator aequorin, demonstrated that the amount of Ca^{2+} released decreased with maintained contractile activity. Additionally, following a prolonged activation pulse, intracellular Ca^{2+} stays above basal ('resting') levels for an extended period of time [31]. All of the above observations can be related to changes in specific steps in E-C coupling.

Muscle activation and force development are regulated by Ca^{2+}. The intracellular Ca^{2+} content depends upon the relative activity of Ca^{2+} release from the SR, and Ca^{2+} removal processes. The latter consists of the activity of the SR Ca^{2+} pump and Ca^{2+} binding proteins, particularly parvalbumin [34]. Calcium release from the SR can be modeled by:

$$F = P \cdot \Delta C, \tag{1}$$

where F = Ca^{2+} flux, P = permeability, and ΔC = Ca^{2+} concentration gradient between SR lumen and intracellular fluid.

P is reasonably well described as the product of independent activation (m) and inactivation (h) variables [48]:

$$P = m(t) \cdot h(t), \tag{2}$$

where t = time.

The activation variable (m) seems to be determined approximately linearly by the quantity of t-tubular charge (Q) moved beyond a threshold (Q_{th}) [35]:

$$P = K[Q(t) - Q_{th}] \cdot h(t), \tag{3}$$

Consequently, Ca^{2+} flux (F) is described by:

$$F = K[Q(t) - Q_{th}] \cdot h(t) \cdot \Delta C, \tag{4}$$

where K = proportionality constant between t-tubular charge movement and the degree of SR Ca^{2+} channel opening.

This last equation permits an organized consideration of E-C coupling factors that may be involved in muscle fatigue. All factors that reduced release flux F will contribute to fatigue, they include, factors that reduce the charge movement term ($Q-Q_{th}$), the intrinsic inactivation term [h(t)], and the concentration gradient (ΔC). Additionally, considering that force is related to intracellular Ca^{2+} which in turn is dependent on the Ca^{2+} release and removal fluxes, any influences that decrease removal of Ca^{2+} will tend to prolong the Ca^{2+} transient and slow relaxation.

Charge Movement Term

It has recently been shown that low extracellular Ca^{2+} reduces intra-membrane t-tubular charge movement (fig. 1 No. 2) and Ca^{2+} release (fig. 1 No. 4) from the SR [41]. Consequently, if t-tubular lumen Ca^{2+} declines with maintained contractile activity, one would predict a reduced t-tubular charge movement (Q), Ca^{2+} release (F), and force output (fatigue). However, Bianchi and Narayan [4] found contractile activity to increase t-tubular lumen Ca^{2+} at a rate directly dependent on the stimulation frequency. An elevated t-tubular lumen Ca^{2+} should increase intramembrane t-tubular charge (Q) and enhance Ca^{2+} release (F) [41].

Intrinsic Inactivation Term

The rate and extent of Ca^{2+} release from the SR could decline during contractile activity even with normal or enhanced t-tubular charge movement, if either the coupling step between the t-tubular and SR membrane (fig. 1 No. 3) or the release process itself (fig. 1 No. 4) were inhibited. Prolonged depolarization pulses (200 ms) have been shown to inhibit peak Ca^{2+} release [47]. The mechanism of inhibition was attributed to Ca^{2+} inhibition of Ca^{2+} release via an increased inactivation (smaller h) of the SR Ca^{2+} release channel, and depletion of SR Ca^{2+} stores [8, 47].

Concentration Term

The concentration gradient (ΔC) depends on the SR Ca^{2+} content; depletion of SR Ca^{2+} decreases ΔC and thus release. The depletion of SR Ca^{2+} during continuous contractile activity is likely due to an increased binding of Ca^{2+} to the intracellular Ca^{2+}-binding proteins parvalbumin and the SR pump. Consequently, during activation free Ca^{2+} would be removed more slowly. Additionally, the SR Ca^{2+} pump rate may be slowed in fatigued muscle fibers [39] further reducing SR Ca^{2+} stores, and decreasing ΔC and Ca^{2+} release (see equation 4). These mechanisms would increase the 'resting' intracellular Ca^{2+} [31]. The net effect is that Ca^{2+} would be displaced from the release pool (SR store) to the removal pool where it is unavailable for release. These events would explain the observed prolongation in the Ca^{2+} transient [8] and twitch duration [52] in fatigued muscle.

It is unknown whether or not these factors contribute to skeletal muscle fatigue in vivo. The exact role that changes in intramembrane t-tubular charge, Ca^{2+} release and removal processes play in fatigue awaits a better understanding of the basic steps in E-C coupling, and how such steps are altered during high frequency contractile activity.

Potential Metabolic Fatigue Agents

High intensity exercise involves an energy demand that exceeds the individual's maximal aerobic power, and thus requires a high level of anaerobic metabolism. Consequently, the muscle high energy phosphates, ATP and phosphocreatine (PC), decrease, while inorganic phosphate (P_i), ADP, lactate, and the H^+ ion all increase as fatigue develops. All of these changes have been suggested as possible fatigue-inducing agents [21, 49], and since the development of the needle biopsy technique [2] each has been extensively studied [21]. To avoid fatigue adequate tissue ATP levels must be maintained as this substrate supplies the immediate source of energy for force generation by the myosin cross-bridges. ATP is also needed in the functioning of the Na/K (fig. 1) pump, which is essential in the maintenance of a normal sarcolemma and t-tubular AP. Additionally, ATP is a substrate of the SR ATPase (fig. 1 No. 5) and thus required in the process of Ca^{2+} reuptake by the SR. As discussed above, a disturbance in any of these processes could lead to muscle fatigue. Although tissue ATP concentration decreases during intense muscular contraction, it does not appear to limit force output or directly cause muscular fatigue. The classic experiments of Karlsson and Saltin [30] perhaps best illustrate the absence of correlated changes of ATP and performance. In this work, the needle biopsy technique was used to evaluate substrate changes following exercise to exhaustion at three different work loads. After 2 min of work, ATP and PC were depleted to the same extent at all loads, while fatigue occurred only with the highest work load. These results are not unequivocal since the biopsy was acquired some seconds after work ceased, and the sample represented an average tissue ATP which might not reflect the concentration existing at the cross-bridges. However, we have also observed a lack of a correlation between ATP and force in isolated muscles studied in vitro [18, 20]. In these experiments, ATP showed complete recovery in the first 15 s following a fatiguing stimulation bout, while peak tetanic tension (P_o) remained depressed.

PC declines with contractile activity [18, 49], and some have suggested that low muscle PC could induce fatigue [49]. However, the declines in PC and tension during contractile activity follow different time courses making a casual relationship unlikely [10, 18]. PC participates in the movement of ATP from the mitochondria to the cross-bridges, a process called the PC-ATP shuttle [46]. The possibility exists that a critically low PC concentration may disrupt this shuttle system and slow the rate of ADP rephosphorylation to ATP. This could lead to a critically low ATP at the cross-bridges and thus produce muscle fatigue. The exact role of PC (if any) in the etiology of muscular fatigue awaits further study of the PC-ATP shuttle mechanism.

Muscular contraction involves the hydrolysis of ATP by the actomyosin ATPase producing energy and yielding ADP, P_i, and the H^+ ion as end products. All three of these products increase during contractile activity [10] and could cause fatigue by a direct inhibition of hydrolysis. Recent studies, utilizing the single-skinned fiber preparation, have shown a reduced peak tension in response to both an elevated H^+ and P_i, while 1 mM ADP increased tension [33, 40].

Dawson et al. [11] have reported a strong linear relationship between the concentration of the acid form of P_i ($H_2PO_4^{-1}$) and the decline in force development in anaerobically stimulated frog gastrocnemius muscle. They suggest that $H_2PO_4^{-1}$ reduces force by product inhibition, and that the effect of H^+ on force production can be attributed to an acidification of P_i. However, the preponderance of evidence suggests that the H^+ ion has direct effects independent of P_i [7, 9, 24, 33, 37, 40]. Muscle fatigue is generally associated with a reduced maximal shortening velocity (V_{max}) [14]. It seems unlikely that increases in $H_2PO_4^{-1}$ can by itself explain

Table I. Skeletal muscle pH

Method	Rest value[1]	Fatigue value	Reference
Muscle homogenate	6.92 ± 0.03	6.41 ± 0.04	Hermansen and Osnes [25]
Calculated from			
$HCO_3^- + PCO_2$	7.04 ± 0.05	6.37 ± 0.11	Sahlin et al. [43]
Micro-electrode	7.07 ± 0.007		Aickin and Thomas [1]
DMO method	7.06 ± 0.02		Roos [42]

[1] Values are means $\pm$ SE except for Sahlin et al., where $\pm$ SD are listed.

fatigue as 30 mM P_i had no effect on V_{max}, while decreasing pH from 7.0 to 6.5 decreased force and V_{max} and increased the free Ca^{2+} required for one-half maximal actomyosin ATPase activity [7, 33]. A decreased free energy change or affinity for the hydrolysis of ATP could mediate fatigue. However, affinity shows only a small decline with contractile activity, and the change appears unrelated to either the reduced force [10] or V_{max} [33].

The H^+ ion is a particularly interesting potential fatigue agent as it could produce fatigue at numerous sites. In addition to a direct inhibition of the actomyosin ATPase and ATP hydrolysis, a build-up in the intracellular H^+ ion (decreased intracellular pH, abbreviated pH_i) could induce fatigue by: (1) inhibiting phosphofructokinase and thus the glycolytic rate [9]; (2) competitive inhibition of Ca^{2+} binding to troponin C reducing cross-bridge activation [7, 24], and (3) inhibiting the SR-ATPase reducing Ca^{2+} reuptake and subsequently Ca^{2+} release [39].

A major source of H^+ ion production during intense muscular activity is the anaerobic production of lactic acid, the majority of which dissociates into lactate and H^+ ions. As early as 1907 lactic acid was implicated as a possible fatigue agent [23]. This hypothesis gained popularity following the work of Hill [27] in the late 1920s. It is now generally thought that fatigue results from the elevated H^+ rather than lactate or the undissociated lactic acid. Resting skeletal muscle pH, determined by a variety of techniques, is approximately 7.00 [21, 37] and declines with intense muscular contraction to values below 6.5 (table I). Utilizing the microelectrode technique, we have recently measured postfatigue muscle pH values as low as 6.33 [37].

At least two observations lend support to the concept that an elevated H^+ ion concentration inhibits glycolysis. Hill [27] found that lactate formation during muscle stimulation stopped when the intracellular pH dropped to 6.3. Secondly, Hermansen and Osnes [25] measured the pH of muscle homogenates and observed no change during a 60-second measurement period for the most acidic homogenates of fatigued muscle, while the pH values of the homogenates from resting muscle showed a marked fall due to significant glycolysis during the measurement period. Sahlin et al. [43] suggest that this inhibition of glycolysis by the H^+ ion may be the limiting factor for performance of intense exercise. However, this conclusion is not supported by the results described above in which the change in ATP and force were not significantly correlated [10, 18, 20]. If the inhibition of glycolysis was causative in fatigue, one would expect the decline in tissue ATP to reach limiting levels. Consequently, a significant correlation between force and ATP would exist during the development of and recovery from fatigue.

Studies on skinned fibers [17] have definitively shown acidosis to depress the force output of skeletal as well as cardiac muscle. Decreasing pH from 7.4 to 6.2 not only reduced the maximal tension generated in the presence of optimal free Ca^{2+}, but also increased the threshold of free Ca^{2+} required for contraction, and shifted the force-PCa curve to the right such that higher free Ca^{2+} was required to reach a given tension [17]. Fast-twitch fibers were found to be more sensitive to the acidotic depression of maximal tension than the slow muscle fibers [12]. These effects may be mediated by a H^+ ion interference with Ca^{2+} binding to troponin [7, 24]. However, the results of Donaldson and Hermansen [12] and Fabiato and Fabiato [17] indicate that the effect is not entirely a simple competitive inhibition of Ca^{2+} binding to troponin. The observed depression of maximal tension was completely reversible on returning to neutral pH but could not be overcome by increasing free Ca^{2+} [17]. Fabiato and Fabiato [17] suggest that the H^+ ion effect may, in part, be acting at some step after Ca^{2+} interaction with troponin, perhaps directly affecting the myosin molecule. A direct effect on myosin is supported by the observation that the maximum rigor tension is reduced by 33% when pH is changed from 7.00 to 6.2.

The rate of ATP hydrolysis by actomyosin is thought to limit the maximal speed of muscle shortening (V_{max}). Consequently, an elevated H^+ ion which inhibits the ATPase should decrease V_{max} as well as P_o. Edman and Mattiazzi [14] have shown that V_{max} indeed decreases with fatigue, but not until peak tension has fallen by at least 10%.

We have recently observed a high negative correlation between free H^+ ion content and P_0 during recovery, a result fully consistent with the H^+ ion theory of muscular fatigue. However, this work [37] and previously published work [20] clearly show force to recover in two phases – a short (~ 30 s) rapid phase followed by a slower relatively prolonged (~ 15 min) phase of recovery. The immediate rapid phase of recovery cannot be explained by the H^+ ion theory since cell pH_i is actually decreasing during this time (probably due to the rapid resynthesis of PC, a H^+ ion-generating reaction). This rapid phase of force recovery is likely explained by the reversal of a non-H^+ ion-mediated alteration in E-C coupling. The second slower phase of recovery shows a high negative correlation with the H^+ ion, and probably results, at least in part, from the removal of the excess intracellular H^+ ion.

Changes in pH may affect Ca^{2+} regulation by disturbing E-C coupling (discussed above) and/or Ca^{2+} reuptake. Nakamura and Schwartz [39] found a decreased pH to increase the Ca^{2+} binding capacity of isolated SR membranes, and suggested that a drop in pH might reduce the amount of Ca^{2+} released from the SR during excitation. Changes in free H^+ ion may also affect the Ca^{2+}-binding properties of the Ca^{2+}-binding protein parvalbumin (a protein found in relatively high concentrations in fast-twitch fibers) which by itself would alter the Ca^{2+} transient and force output.

A major challenge to the field of exercise science is to establish which (if any) of the above-mentioned H^+ ion effects actually occurs in vivo, and determine the relative importance of increased H^+ and P_i and alterations in E-C coupling in the fatigue process.

Frequency Dependent Fatigue

Edwards et al. [15] studying man and Jones et al. [29] evaluating animal as well as human muscle observed that fatigue in response to low frequencies of stimulation persisted after force in response to high frequencies had fully recovered. Since skeletal muscle is generally activated by low frequency (10–30 Hz) stimulation [6, 15], this selective effect could have important functional implications. Edwards et al. [15] referred to this selective effect as low frequency fatigue and suggested that it was mediated by disturbances in E-C coupling. Reduction in force at low frequencies means that the force-frequency relationship is shifted to the right [see figure 7, ref. 15]. This result is difficult to understand considering that muscle contractile duration is prolonged after fatigue [52], and thus force production at low frequencies should benefit from the resultant increase in fusion.

Recently, we reexamined the effect of high- and low-frequency stimulation on force recovery at both ends of the force-frequency relationship [38]. We observed force to recover more rapidly at low compared to high Hz stimulation. Furthermore, in the fatigued state the muscle generated considerably more force during low (5 Hz) compared to high (75 Hz) frequency stimulation [see figure 10, ref. 38]. Jones et al. [28] observed a similar result [see figure 2, ref. 28], and suggested that in vivo the activation frequency during the course of a sustained voluntary contraction declined, thus taking advantage of the relatively higher forces at low frequencies as fatigue developed. The higher force at low frequencies in fatigued muscle is in part due to the increased fusion that results from the prolonged contractile duration [see figure 3, ref. 38]. Thus, in our experiments, the force-frequency relation shifted to the left, and, correspondingly, the optimal stimulation frequency for peak force production declined [38]. Thus, we have not been able to confirm the observation of Edwards et al. [15] demonstrating persistent low frequency fatigue. Our results (utilizing an in vitro muscle preparation) clearly show force production at low frequency to recover more rapidly than at high frequency after fatigue produced by either high- or low-frequency stimulation [38].

Endurance Exercise

Numerous factors have been linked to fatigue resulting from prolonged endurance activity. These include depletion of muscle and liver glycogen, decreases in blood glucose, dehydration, and increases in body temperature [21]. Undoubtedly, each of these factors contributes to fatigue to a varying degree, their relative importance depending on the environmental conditions and the nature of the activity. This section will review some of these potential fatigue factors, particularly carbohydrate depletion, and present evidence linking an alteration in SR function to the development of fatigue during prolonged exercise.

Glycogen Depletion

In 1896, Chauveau suggested that the rate of carbohydrate utilization was dependent on the intensity of work [21]. This belief was based on the observation that the respiratory exchange ratio (RQ) increased from 0.75 during rest to 0.95 during exercise. With the development of the needle biopsy technique [2], these early theories based on RQ were proven correct by direct measurements of glycogen utilization at different work intensities [44]. Glycogen utilization was found to increase from 0.3 to 3.4 glucose units $\times$ kg^{-1} $\times$ min^{-1} as the relative work load increased from 25 to 100% of the maximal oxygen uptake VO_{2max} [44]. Muscle glycogen depletion coincided with exhaustion during prolonged work bouts requiring approximately 75% of VO_{2max}. With work loads below 50 or above 90% of VO_{2max}, ample muscle glycogen remained at exhaustion [44]. The rate of body carbohydrate usage is dependent not only on the intensity of the work but also on the individual's state of fitness [44]. At a given work load, trained individuals have a lower RQ and deplete glycogen at a slower rate than untrained men [44]. The observation that the trained individual can work longer supports the hypothesis that depletion of body carbohydrate stores is not only correlated with, but is causative of, muscular fatigue during endurance activity. The exact mechanism of this protective effect is unknown. Although muscle glycogen represents an important fuel source, adequate levels of free fatty acids (FFA) and in most cases blood glucose are available at exhaustion. One possibility is that a certain level of muscle glycogen metabolism is required for either the optimal production of NADH and electron transport or the maintenance of fat oxidation, perhaps intermediates of the Krebs cycle become limiting without adequate glycogen metabolism. Alternatively, the translocation of FFA into mitochondria may be rate limiting and/or a high concentration of long-chain FFA might inhibit ATP translocation across the mitochondria membrane. It seems apparent that future work will have to focus on the mechanisms by which glycogen depletion alters muscle function.

Other Factors

It seems unlikely that glycogen depletion is an exclusive fatigue factor during endurance exercise. Other potential candidates include disruption of important intracellular organelles such as the mitochondria, the SR or the myofilaments. Significant mitochondrial damage seems unlikely as electron-microscopic analysis of fatigued muscle demonstrates a normal mitochondrial structure, and their capacity to oxidize substrate and generate ATP is unchanged by exercise to exhaustion [50].

The contractile proteins and, in particular, the myofibril ATPase (turnover measured by V_{max}) appear relatively resistant to fatigue. Following a prolonged swim, the myofibril Mg^{2+} ATPase of fast and slow rat hindlimb muscles was unaltered (table II). The V_{max} of the slow soleus

Table II. Effect of endurance swim to exhaustion on myofibril ATPase, V_{max} and P_o

		Myofibril ATPase µmol/mg/min	V_{max} fiber lengths/s	P_o g/cm^2
Slow soleus	Control	0.128 ± 0.008	2.9 ± 0.2	2,482 ± 208
	Fatigued	0.104 ± 0.011	2.7 ± 0.2	1,844 ± 203*
Fast EDL	Control	0.304 ± 0.031	7.6 ± 0.8	2,397 ± 171
	Fatigued	0.309 ± 0.030	5.2 ± 0.7*	633 ± 190*
Fast SVL	Control	0.301 ± 0.021	9.5 ± 1.2	1,757 ± 88
	Fatigued	0.343 ± 0.021	9.8 ± 0.8	1,713 ± 86

Values represent means ± SE.
Significantly different * $p < 0.05$.

showed no significant change despite a 26% decline in peak tension. The fast-twitch extensor digitorum longus (EDL) did undergo a significant 34% decline in V_{max}, but this muscle exhibited extreme fatigue with the P_o falling by more than 70% (table II) [22]. In this case the fatigued (inactive) fibers might have provided a significant internal drag during the unloaded and lightly loaded contractions. These results imply that the activity of the myofibril ATPase and its functional correlate V_{max} are relatively resistant to alteration during prolonged exercise.

In the same study, we evaluated the functional capacity of isolated SR membranes [22]. The SR is an intracellular membrane system primarily involved in the regulation of intracellular Ca^{2+}. Alteration in the force transient of contraction (a reflection of an altered Ca^{2+} transient) is a common observation with fatigue and thus the SR may be involved in the etiology of muscular fatigue. Our results showed that the prolonged swim had no effect on the amount of SR isolated (mg/g tissue) from any of the fast muscles, but produced a significant decrease in SR protein isolated from the slow soleus (0.81 ± 0.05 vs. 0.57 ± 0.05 mg/g, control vs. fatigued). This decrease is unexplained but could reflect an elevated proteolytic enzyme activity shown to exist in fatigued muscle [54]. None of the muscles studied exhibited any change in the SR Ca^{2+}-stimulated ATPase activity. However, Ca^{2+} uptake by the SR vesicles (µmol·mg^{-1} SR) was depressed in the slow soleus and fast-twitch red region of the vastus lateralis. A decreased Ca^{2+} uptake with no change in the SR ATPase activity suggests either an uncoupling of the transport or a 'leaky' membrane allowing Ca^{2+} flux back into the intracellular fluid. Although our results clearly show a major change in the SR, more experiments studying the effects of prolonged activity on the kinetic properties of Ca^{2+} uptake and release during excitation are required before the exact nature of this change and its effect on muscle function can be elucidated.

The prolonged swim produced a significant decrease in glycogen concentration in muscles representative of the slow type I, fast type IIa, and fast type IIb fiber. Interestingly, the type IIb muscle showed no fatigue as reflected by an unaltered P_o. Furthermore, despite significant glycogen depletion, the fast type IIb muscle showed no change in any of the contractile or biochemical properties measured [22]. The apparent explanation is that the type IIb (fast white glycolytic) fiber was recruited less frequently during the endurance activity and that this fiber's heavy reliance on glycolysis produced similar levels of muscle glycogen usage as the other fiber types despite fewer total contractions. It is apparent from these results that muscle fatigue during endurance activity is related in some way to the degree of muscle usage, and not entirely dependent on the extent of glycogen depletion. An important unanswered question is whether or not glycogen depletion somehow mediates (and hence is a prerequisite for) the disruption of intracellular organelles such as the SR.

The observed disruption in protein systems such as the SR coupled with an increased concentration of intracellular metabolites would be expected to increase the intracellular solute concentration and tissue water. An elevated tissue water would produce swelling and thus could lead to muscle soreness. This possibility is supported by structural studies relating muscle soreness to changes in various intracellular organelles [26]. The time course of recovery from muscle soreness (days) exceeds that observed with fatigue (minutes), and reflects the time required to synthesize new muscle protein.

In conclusion, the studies described in this chapter illustrate the complex nature of muscle fatigue. Following short duration, high intensity exercise, recovery in force production generally shows two components likely caused by separate mechanisms: (1) a rapidly reversible non-H^+ ion (or P_i) mediated perturbation in E-C coupling and Ca^{2+} regulation, and (2) a slower change likely involving several sites and steps in muscle contraction, and mediated at least in part by H^+ and P_i. The potential mechanisms of the deleterious effects of the H^+ and P_i have been described.

In prolonged endurance exercise, the depletion of body carbohydrate stores frequently occurs, and muscle glycogen depletion likely represents one important fatigue agent. Undoubtedly, other factors are involved as muscle glycogen depletion can exist without fatigue. In addition to glycogen depletion, muscle organelles, particularly the SR, are probably involved in the fatigue process.

References

1 Aickin, C.C.; Thomas, R.C.: Microelectrode measurement of the intracellular pH and buffering power of mouse soleus muscle fibers. J. Physiol., Lond. *267:* 791–810 (1977).
2 Bergtrom, J.: Muscle electrolytes in man. Scand. J. clin. Lab. Invest. *68:* suppl. (1962).
3 Bezanilla, F.; Caputo, C.; Gonzalez-Serratos, H.; Venosa, R.A.: Sodium dependence of the inward spread of activation in isolated twitch muscle fibers of the frog. J. Physiol., Lond. *223:* 507–523 (1972).
4 Bianchi, C.P.; Narayan, S.: Muscle fatigue and the role of transverse tubules. Science *215:* 295–296 (1982).
5 Bigland-Ritchie, B.; Jones, D.A.; Woods, J.J.: Excitation frequency and muscle fatigue: electrical responses during human voluntary and stimulated contractions. Expl Neurol. *64:* 414–427 (1979).
6 Bigland-Ritchie, B.; Dawson, N.J.; Johansson, R.S.; Leppold, O.C.J.: Reflex origin for the slowing of motoneurone firing rates in fatigue of human voluntary contractions. J. Physiol., Lond. *379:* 451–459 (1986).
7 Blanchard, E.M.; Pan, B.S.; Solaro, R.J.: The effect of acidic pH on the ATPase activity and troponin Ca^{2+} binding of rabbit skeletal myofilaments. J. biol. Chem. *259:* 3181–3186 (1984).
8 Blinks, J.R.; Rudel, R.; Taylor, S.R.: Calcium transients in isolated amphibian skeletal muscle fibres: detection with aequorin. J. Physiol., Lond. *277:* 291–323 (1978).
9 Danforth, W.H.: Activation of glycolytic pathway in muscle; in Chance, Control of energy metabolites, pp. 287–297 (Academic Press, New York 1965).
10 Dawson, M.J.; Gadian, D.G.; Wilkie, D.R.: Muscular fatigue investigated by phosphorus nuclear magnetic resonance. Nature, Lond. *274:* 861–866 (1978).
11 Dawson, M.J.; Smith, S.; Wilkie, D.R.: The $[H_2PO_4^{-1}]$ may determine cross-bridge cycling rate and force production in living fatiguing muscle. Biophys. J. *49:* 268a (1986).
12 Donaldson, S.K.B.; Hermansen, L.: Differential, direct effects of H^+ on Ca^{2+}-activated force of skinned fibers from the soleus, cardiac and adductor magnus muscles of rabbits. Pflügers Arch. *376:* 55–65 (1978).

13 Eberstein, A.; Sandow, A.: Fatigue mechanisms in muscle fibers; in Gutmann, The effect of use and disuse on neuromuscular function, pp. 515–526 (Academic Publishing House of Czechoslovak Academy of Science, Prague 1963).

14 Edman, K.A.P.; Mattiazzi, A.R.: Effects of fatigue and altered pH on isometric force and velocity of shortening at zero load in frog muscle fibers. J. mus. Res. Cell Mot. 2: 321–334 (1981).

15 Edwards, R.H.T.; Hill, D.K.; Jones, D.A.; Merton, P.A.: Fatigue of long duration in human skeletal muscle after exercise. J. Physiol., Lond. 272: 769–778 (1977).

16 Edwards, R.H.T.: Human muscle function and fatigue; in Porter, Human muscle fatigue: physiological mechanisms, pp. 1–18 (Pitman, London 1981).

17 Fabiato, A.; Fabiato, F.: Effects of pH on the myofilaments and the sarcoplasmic reticulum of skinned cells from cardiac and skeletal muscles. J. Physiol., Lond. 276: 233–255 (1978).

18 Fitts, R.H.; Holloszy, J.O.: Lactate and contractile force in frog muscle during development of fatigue and recovery. Am. J. Physiol. 231: 430–433 (1976).

19 Fitts, R.H.; Holloszy, J.O.: Contractile properties of rat soleus muscle: effects of training and fatigue. Am. J. Physiol. 2: C86–C91 (1977).

20 Fitts, R.H.; Holloszy, J.O.: Effects of fatigue and recovery on contractile properties of frog muscle. J. appl. Physiol. 45: 899–902 (1978).

21 Fitts, R.H.; Kim, D.H.; Witzmann, F.A.: The development of fatigue during high intensity and endurance exercise; in Nagle, Exercise in health and disease, pp. 118–135 (Thomas, Springfield 1981).

22 Fitts, R.H.; Courtright, J.B.; Kim, D.H.; Witzmann, F.A.: Muscle fatigue with prolonged exercise: contractile and biochemical alterations. Am. J. Physiol. 242: C65–C73 (1982).

23 Fletcher, W.W.; Hopkins, F.G.: Lactic acid in mammalian muscle. J. Physiol., Lond. 35: 247–303 (1907).

24 Fuchs, F.; Reddy, Y.; Briggs, F.N.: The interaction of cations with the calcium-binding site of troponin. Biochim. biophys. Acta 221: 407–409 (1970).

25 Hermansen, L.; Osnes, J.: Blood and muscle pH after maximal exercise in man. J. appl. Physiol. 32: 304–308 (1972).

26 Hikida, R.S.; Staron, R.S.; Hagerman, F.C.; Sherman, W.M.; Costill, S.L.: Muscle fiber necrosis associated with human marathon runners. J. neurol. Sci. 59: 185–203 (1983).

27 Hill, A.V.: The absolute value of the isometric heat coefficient T1/H in a muscle twitch, and the effect of stimulation and fatigue. Proc. R. Soc., Lond., ser. B 103: 163–170 (1928).

28 Jones, D.A.; Bigland-Ritchie, B.; Edwards, R.H.T.: Excitation frequency and muscle fatigue. Mechanical responses during voluntary and stimulated contractions. Expl Neurol. 64: 401–413 (1979).

29 Jones, D.A.; Howell, S.; Roussos, C.; Edwards, R.H.T.: Low-frequency fatigue in isolated skeletal muscles and the effects of methylxanthines. Clin. Sci. 63: 161–167 (1982).

30 Karlsson, J.; Saltin, B.: Lactate, ATP, and CP in working muscles during exhaustive exercise in man. J. appl. Physiol. 29: 598–602 (1970).

31 Klein, M.G.; Simon, B.J.; Szucs, G.; Schneider, M.F.: Myoplasmic calcium tran-

sients monitored simultaneously with high and low affinity calcium indicators. Biophys. J. *51:* 199a (1987).

32 Krnjevic, K.; Miledi, R.: Failure of neuromuscular propagation in rats. J. Physiol., Lond. *140:* 440–461 (1958).

33 Luney, D.J.E.; Godt, R.D.: The effects of pH, ADP, inorganic phosphate (P_i), and affinity on the maximum velocity of shortening and force production of skinned rabbit muscle fibers. Biophys. J. *51:* 468a (1987).

34 Melzner, W.; Rios, E.; Schneider, M.F.: The removal of myoplasmic free calcium following calcium release in frog skeletal muscle. J. Physiol., Lond. *372:* 261–292 (1986).

35 Melzer, W.; Schneider, M.F.; Simon, B.J.; Szucs, G.: Intramembrane charge movement and calcium release in frog skeletal muscle. J. Physiol., Lond. *373:* 481–511 (1986).

36 Metzger, J.M.; Fitts, R.H.: Fatigue from high- and low-frequency muscle stimulation: Role of sarcolemma action potentials. Expl Neurol. *93:* 320–333 (1986).

37 Metzger, J.M.; Fitts, R.H.: Role of intracellular pH in muscle fatigue. J. appl. Physiol. *62:* 1392–1397 (1987).

38 Metzger, J.M.; Fitts, R.H.: Fatigue from high- and low-frequency muscle stimulation: contractile and biochemical alterations. J. appl. Physiol. *62:* 2075–2082 (1987).

39 Nakamura, Y.; Schwartz, A.: The influence of hydrogen ion concentration on calcium binding and release by skeletal muscle sarcoplasmic reticulum. J. gen. Physiol. *59:* 22–32 (1971).

40 Nosek, T.M.; Fender, K.Y.; Godt, R.E.: It is diprotonated inorganic phosphate that depresses force in skinned skeletal muscle fibers. Science *236:* 191–193 (1987).

41 Pizarro, G.; Fitts, R.; Brum, G.; Rodriguez, M.; Rios, E.: Simultaneous recovery of charge movement and Ca release in skeletal muscle. Biophys. *51:* 101a (1987).

42 Roos, A.; Boron, W.F.: Intracellular pH transients in rat diaphragm muscle measured with DMO. Am. J. Physiol. *235:* C49–C54 (1978).

43 Sahlin, K.; Harris, R.C.; Nylind, B.; Hultman, E.: Lactate content and pH in muscle samples obtained after dynamic exercise. Pflügers Arch. *367:* 143–149 (1976).

44 Saltin, B.; Karlsson, J.: Muscle glycogen utilization during work of different intensities; in Pernow, Muscle metabolism during exercise, pp. 289–299 (Plenum Press, New York 1971).

45 Sandow, A.: Excitation-contraction coupling in muscular response. Yale J. biol. Med. *25:* 176–201 (1952).

46 Savabi, F.; Geiger, P.J.; Bessman, S.P.: Myofibrillar end of the creatine phosphate energy shuttle. Am. J. Physiol. *247:* C424–C432 (1984).

47 Simon, B.J.; Schneider, M.F.; Szucs, G.: Inactivation of sarcoplasmic reticulum calcium release in frog skeletal muscle. J. gen. Physiol. *86:* 36a (1985).

48 Simon, B.J.; Schneider, M.F.: A comparison of the kinetics of charge movement and activation of SR calcium release during excitation in frog skeletal muscle. Biophys. J. *51:* 550a (1987).

49 Simonson, E.: Accumulation of metabolites; in Simonson, Physiology of work capacity and fatigue, pp. 9–25 (Thomas, Springfield 1971).

50 Terjung, R.L.; Baldwin, K.M.; Mole, P.A.; Klinkerfuss, G.H.; Holloszy, J.O.: Effect of running to exhaustion on skeletal muscle mitochondria: a biochemical study. Am. J. Physiol. *223:* 549–554 (1972).

51   Vander, A.J.; Sherman, J.H.; Luciano, D.S.: Chapter 10: muscle; in Human physi-
      ology: the mechanisms of body function; 3rd ed., pp. 211–249 (McGraw-Hill, New
      York 1980).
52   Vergara, J.L.; Rapoport, S.I.; Nassar-Gentiria, V.: Fatigue and posttetanic potentia-
      tion in single muscle fibers of the frog. Am. J. Physiol. *1:* C185–C190 (1977).
53   Vergara, J.; Tsien, R.Y.; Delay, M.: Inositol 1,4,5-triphosphate. A possible chemical
      link in excitation-contraction coupling in muscle. Proc. natn. Acad. Sci. USA *82:*
      6352–6356 (1985).
54   Vihko, V.; Salminen, A.; Rantamaki, J.: Exhaustive exercise, endurance training,
      and acid hydrolase activity in skeletal muscle. J. appl. Physiol. *47:* 43–50 (1979).

Dr. Robert H. Fitts, Department of Biology, Marquette University,
Milwaukee, WI 53233 (USA)

Poortmans JR (ed): Principles of Exercise Biochemistry.
Med Sport Sci. Basel, Karger, 1988, vol 27, pp 230–253.

Exercise and Metabolic Disorders

John C. Young, Neil B. Ruderman

Department of Health Sciences, Sargent College, Boston University, and Division
of Diabetes and Metabolism, University Hospital and BUMC, Boston, Mass., USA

Introduction

It has long been appreciated that exercise can have profound effects in
patients with metabolic disorders. Thus, before the isolation of insulin by
Banting and Best in 1921, exercise was considered a useful means of
diminishing hyperglycemia in certain patients with diabetes. Likewise,
exercise has been prescribed as an adjunct to diet in the treatment of obesity. This chapter will review present-day concepts of how exercise impacts
on these and other metabolic disorders. The emphasis will be on patients
with diabetes mellitus, since it is in these individuals that the therapeutic
role of exercise has been most extensively examined. In addition, the role
of exercise in the therapy and/or prevention of obesity, osteoporosis and
certain disturbances of lipoprotein metabolism in the diabetic will be discussed.

Diabetes mellitus and Impaired Glucose Tolerance

Classification
The two major subtypes of diabetes are described in table I. It has
become increasingly clear in recent years that entities previously referred
to as juvenile- and maturity-onset diabetes are genetically distinct disorders that resemble each other because of the presence of hyperglycemia
and a propensity to vascular and neurological complications [31].

Type 1 (juvenile-onset) or insulin-dependent diabetes is characterized by an absolute deficiency of insulin as a result of autoimmune destruction of the insulin-producing beta-cells of the islets of Langerhans. Patients require exogenous insulin by injection to sustain life. Without it they become profoundly hyperglycemic and rapidly break down the body's stores of protein and fat. Clinically, this situation is referred to as diabetic ketoacidosis because of the marked decrease in plasma pH and an increase in the circulation of two lipid-derived substances, β-hydroxybutaric and acetoacetic acid, the ketone bodies. As will be discussed in detail, exercise can present special problems in patients with type 1 diabetes because of hormone-fuel imbalances.

Type 2 (maturity-onset) or non-insulin-dependent diabetes accounts for upwards of 85% of all diabetics in the western world. Most patients are usually above age 35 and they are frequently obese. The precise etiology of type 2 diabetes is unknown; however, in contrast to type 1 diabetes, it does not appear to involve autoimmune destruction of the beta-cells. Patients with type 2 diabetes can develop the same vascular and neurological complications as their type 1 counterparts, but except under special circumstances, e.g. severe stress, they do not develop ketoacidosis. Insulin resistance and, in some individuals, insulin deficiency of a milder degree than is found in the type 1 diabetic characterize this patient group. Standard therapy entails the use of diet to diminish adiposity, frequently oral agents, the sulfonylureas, and, in a significant percentage of patients, exogenous insulin. It is in the type 2 diabetic that the therapeutic role of exercise has been most extensively examined.

Table I. Characteristics of patients with different types of diabetes[1]

	Insulin-dependent diabetes	Non-insulin-dependent diabetes
Synonyms	type 1	type 2
Old name	juvenile onset	maturity onset
Age at onset	< 35	> 35
Insulin deficiency	absolute	relative
Autoimmune origin	yes	probably no
Macrovascular disease	++	+++
Microvascular disease	++++	++

[1] A third major group we will discuss will be patients with impaired glucose tolerance.

A third group of patients with hyperglycemia have a normal fasting blood glucose level, but higher than normal levels when challenged with a glucose load. Such patients were previously referred to as chemical diabetics but are now classified as glucose intolerant. This group is noteworthy for two reasons: first, approximately one-third of them eventually develop type 2 diabetes, and, second, they appear to have a similar predilection to premature atherosclerosis, although not to microvascular disease, as do patients with overt diabetes. In this group, as in the type 2 diabetic, exercise may play a therapeutic role.

Type 1 Diabetes

Hormone-Fuel Interrelationships during Exercise in Normal Humans

During exercise, muscle utilizes metabolic fuels at an increased rate to provide the energy required for contraction. As reviewed elsewhere in this volume, the type of fuel utilized is dependent on a number of factors including the intensity and duration of the exercise, and the nutritional state of the individual. Thus, in normal humans, muscle glycogen is the predominant fuel during very strenuous, short-duration exercise, whereas, blood-borne glucose and free fatty acids (FFA) derived from adipose tissue triglyceride are used preferentially during prolonged exercise of low-to-moderate intensity [24]. Likewise, lipid fuels are used in preference to glucose and muscle glycogen in subjects who have fasted or have been consuming a high-fat diet [1].

Fuel usage varies with the duration of exercise, independently of intensity [1, 25]. For instance, during exercise of moderate intensity intramuscular glycogen and triglycerides are initially the primary fuels of muscle (fig. 1). As exercise continues, blood glucose released into the circulation by the liver or absorbed from the gut, and eventually FFA become more important. This switch from local to circulating fuels and from carbohydrates to lipids is important for endurance exercise, since local fuels are limited and fat reserves are much more abundant than carbohydrates. A total substitution of lipid for carbohydrate does not take place, however.

Glucose uptake by muscle increases 4- to 5-fold or more during exercise. Despite this, the concentration of glucose in plasma is maintained, as a result of enhanced glucose production by the liver. This is of special importance, for except under special circumstances, e.g. prolonged starvation, glucose is the sole fuel of brain in exercising humans as it is in humans at rest. The concentration of glucose is maintained during exercise by an

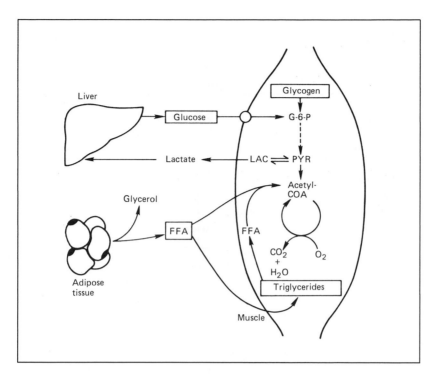

Fig. 1. A schematic diagram of the muscle cell and its intracellular and extracellular fuel sources. During exercise the use of both lipid and carbohydrate fuels increases and nearly all steps in the pathway of glucose metabolism are enhanced. Some of the glucose not oxidized by muscle is released into the circulation as lactate which, in turn, may be used for glucose synthesis by the liver (Cori cycle). Glycerol derived from the hydrolysis of adipose tissue triglyceride may also be used by the liver for gluconeogenesis. Adapted from Richter et al. [41]. Reprinted with permission of *American Journal of Medicine.*

increase in hepatic glucose output. At the onset of exercise this is predominantly due to accelerated glycogenolysis [1]. Hepatic glycogen stores, 50–100 g after an overnight fast, are limited, however, and as exercise continues, gluconeogenesis, the synthesis of glucose from lactate, amino acids, glycerol and pyruvate, plays a more important role [1]. The absolute rate of gluconeogenesis increases 3-fold during prolonged exercise of mild intensity [1]. Hypoglycemia sometimes occurs in normal humans during prolonged exercise, e.g. marathon runners, when hepatic glycogen reserves are depleted and the utilization of glucose by the peripheral tissues exceeds the capacity of the liver for gluconeogenesis.

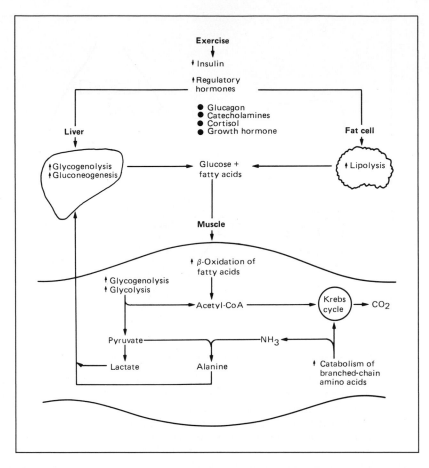

Fig. 2. Schematic diagram shows how hormones mobilized during exercise influence fuel supply to the muscles and outlines the general metabolism of fuel within muscle tissue. Exercise suppresses insulin secretion and stimulates secretion of glucagon, catecholamines, cortisol, and growth hormone. Adapted from Frontera and Adams [15]. Reprinted with permission of *The Physician and Sports Medicine.*

Fuel homeostasis during exercise is modulated by the interplay between insulin on the one hand and neurotransmitters such as norepinephrine and the counterregulatory hormones, epinephrine, glucagon, cortisol and growth hormone, on the other. The overall response of the nervous and endocrine systems to exercise is geared toward increasing hepatic glucose production via glycogenolysis and gluconeogenesis, mobilizing free fatty acids from

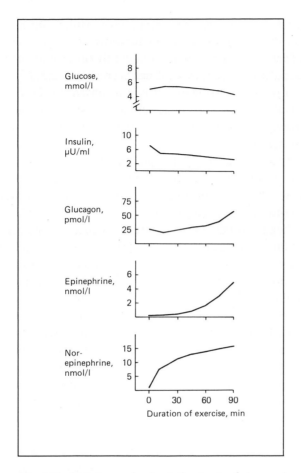

Fig. 3. Typical changes in circulating levels of plasma glucose, insulin, glucagon and the catecholamines during prolonged exercise after an overnight fast. The exercise is of moderate intensity (approximately 60–70% of maximum aerobic capacity). Adapted from Richter et al. [41]. Reprinted with permission of *American Journal of Medicine.*

adipose tissue depots and to some extent stimulating glycogen degradation in muscle (fig. 2). The magnitude of the neural and hormonal responses is dependent on the intensity of the exercise and is diminished after exercise training (see below). Typical changes in plasma levels of insulin, glucagon, and catecholamines during 90 min of moderately intense (60–70% of maximal capacity) exercise are shown in figure 3.

Under normal circumstances, the concentration of insulin diminishes rather early in exercise, due to alpha-adrenergic inhibition of insulin secretion [16]. While this decrease was originally thought to be responsible for the enhancement of hepatic glucose release, it has been shown that hepatic glucose production can increase during exercise in its absence [62]. Possibly, the decrease in insulin serves to facilitate adipose tissue lipolysis (insulin is a potent antilipolytic hormone) or to restrict the use of glucose by nonexercising skeletal muscle.

Hepatic glucose production during prolonged exercise is stimulated by the counterregulatory hormones, glucagon, epinephrine, and cortisol. Plasma glucagon enhances both hepatic glycogenolysis and gluconeogenesis and its concentration tends to rise during prolonged exercise, as the plasma glucose concentration begins to fall. There is some evidence that glucagon also contributes to the increase in hepatic glucose production early in exercise [39, 59]. If so, its effect must be permissive, as the concentration of glucagon in plasma is not increased at this time (fig. 3). Epinephrine has similar, but less-pronounced, effects on hepatic glucose output as does glucagon. Its concentration, like that of glucagon, increases during prolonged exercise in response to a decrease in blood glucose, and this probably is a backup mechanism to maintain normoglycemia. In addition to enhancing hepatic glucose production, epinephrine may act to maintain normoglycemia during exercise by increasing muscle glycogenolysis and lipolysis in adipose tissue, thereby diminishing the need for blood-borne glucose [41]. Plasma cortisol and growth hormone also increase during prolonged exercise; however, their role in regulating fuel homeostasis is less clear.

While decreases in plasma insulin and increases in the counterregulatory hormones stimulate hepatic glucose output during prolonged exercise, it is unlikely that they are responsible for the initial increase in hepatic glucose production. Glucose release by the liver is increased very early in exercise, before the levels of insulin, glucagon, and epinephrine are altered. Presumably, norepinephrine released from sympathetic nerve endings modulates the initial release of glucose by the liver; however, definitive proof of this is lacking.

Altered Hormone-Fuel Interrelations during Exercise in Type 1 Diabetes

Poorly Controlled Patients. This metabolic and hormonal response to exercise as described above in normal humans may be quite different in an insulin-dependent diabetic patient (fig. 4). In a moderately well-controlled

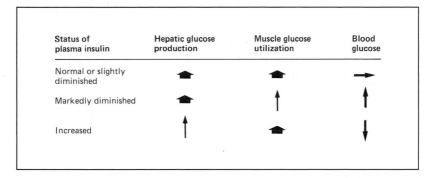

Status of plasma insulin	Hepatic glucose production	Muscle glucose utilization	Blood glucose
Normal or slightly diminished	⬆	⬆	→
Markedly diminished	⬆	↑	↑
Increased	↑	⬆	↓

Fig. 4. Modified from schema of Vranic and Berger [56]. The increase in blood glucose in the insulin-deficient state is probably attenuated by an increased loss of glucose in urine. Adapted from Richter et al. [41]. Reprinted with permission of *American Journal of Medicine.*

type 1 diabetic with some insulin in the circulation, exercise usually causes a modest decrease in plasma glucose, and alterations in plasma FFA, ketone bodies, and most hormones similar to those of normal humans (fig. 5). When these patients exercise regularly, their requirement for insulin may decrease. In contrast, poorly controlled patients usually experience increases in plasma glucose, FFA, and ketone bodies during exercise (fig. 5). This is assumed to be due to initially elevated levels of the counterregulatory hormones that increase further during exercise, and the presence of a low concentration of insulin [6]. In such individuals, hyperglycemia is accentuated because hepatic glucose production which is increased even prior to exercise, increases still further, and at the same time, glucose uptake by skeletal muscle is impaired [7, 57, 58].

The cause of the impairment in glucose uptake by skeletal muscle in poorly controlled patients with diabetes is incompletely understood. Possible factors include high circulating concentrations of FFA, ketone bodies, and epinephrine, and chronic changes in the number of glucose transporters [6, 13, 58]. Lipolysis in adipose tissue is also enhanced by the relative imbalance between insulin and counter-insulin hormones. As a result, plasma FFA levels increase and, when the rate of beta-oxidation of FFA in the liver exceeds the capacity of the Kreb's cycle to utilize acetyl-CoA, ketone bodies are produced. In some patients, exercise may precipitate ketoacidosis. For these reasons, vigorous and prolonged exercise should be avoided by type 1 diabetics who are not adequately insulinized.

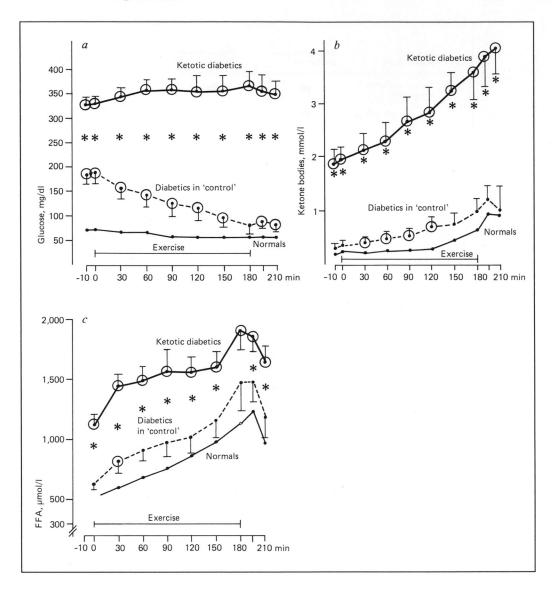

Fig. 5. Effect of prolonged exercise on blood glucose (*a*), plasma ketone bodies (acetoacetate and 3-hydroxybutyrate) (*b*), and plasma FFA (*c*) in healthy control subjects, diabetic patients in moderate metabolic control, and in ketotic diabetic patients. Encircled values are significantly different from corresponding values of the control group (p < 0.05). Stars indicate statistically significant differences (p < 0.05) between corresponding values of the two groups of diabetic patients. Adapted from Berger et al. [6]. Reprinted with permission of *Diabetologia*.

Hypoglycemia. Although exercise causes a modest decrease in plasma glucose in most type 1 diabetics, in some it causes a more marked decrease leading to symptomatic hypoglycemia [28, 41, 44, 56]. The cause of the decrease in glucose in both groups appears to be inappropriately high insulin levels. Plasma insulin is usually higher in the type 1 diabetic than in the nondiabetic during exercise because of increased absorption from an injection site, high insulin levels from a previous injection, or both. Such high insulin levels suppress the increase in hepatic glucose production caused by exercise [6, 16, 39]. Since glucose utilization by working muscles continues at an increased rate, the concentration of glucose in plasma falls. The symptoms of hypoglycemia are generally adrenergic, i.e. tachycardia, sweating, anxiety, and are due to a catecholamine discharge. When blood glucose falls to low levels, and particularly in patients who have a defective adrenergic response, the symptoms will be those of neuroglycopenia. These include confusion, inability to concentrate, lethargy, seizures, and coma.

Clinically significant hypoglycemia is most likely to occur when the exercise is prolonged and/or intense, when the plasma glucose level prior to physical activity is near normal, and when the exercise takes place shortly after the injection of insulin into an exercising limb [28, 44]. Conversely, exercise after a meal is less likely to result in hypoglycemia because the glucose absorption from the gastrointestinal tract is not insulin dependent. A special problem in some type 1 diabetes is an impaired response to hypoglycemia. The glucagon response to hypoglycemia is diminished in many type 1 diabetics, as may be the epinephrine response, for reasons not entirely clear [11, 21]. Exercise-induced hypoglycemia may be particularly dangerous in some type 1 diabetics with autonomic neuropathy, who lack the typical adrenergic warning signs and symptoms, and are more likely to present with neuroglycopenic symptoms.

It is important for the type 1 diabetic who exercises to be aware of factors that predispose to hypoglycemia, and to know how to prevent them. The patient should also be aware that hypoglycemia can occur many hours after the exercise. Hypoglycemia at this time is probably related to a number of factors including the rapid repletion of muscle glycogen stores [17], an associated increase in insulin sensitivity in the previously exercised muscle [40], and a continued suppression of hepatic glucose production by inappropriately high levels of insulin.

The principle means available to the patient for preventing exercise-induced hypoglycemia are (1) ingestion of carbohydrate, and (2) diminution of insulin dose. Ingestion of a snack containing 40–60 g of carbohydrate prior to exercise usually suffices to prevent hypoglycemia during exercise of moderate intensity and duration. When exercise is intense and prolonged, a decrease in the insulin dose is usually necessary. In some patients who develop nocturnal hypoglycemia following exercise late in the day, neither additional carbohydrate ingestion nor a decrease in the insulin dose may be sufficient. In such instances, it may be necessary for the patient to exercise earlier in the day.

As the response of patients to exercise is not uniform, the approach to preventing hypoglycemia both during and after exercise must be individualized. Self-monitoring of blood glucose is particularly helpful in this regard. The effects of carbohydrate ingestion, decrease of insulin dose, and timing of exercise relative to meals must be worked out by each patient on the basis of such blood glucose measurements.

Initial studies suggested that shifting the site of insulin injection from the thigh to the abdomen or arm may prevent hypoglycemia in runners, cyclists, or others undergoing exercise in which the lower extremities are predominantly used [29]. Subsequent investigations suggested that this approach does not work under most conditions, however [27, 53].

Type 2 Diabetes

There are two principal reasons for considering exercise in the treatment of type 2 diabetes: first, in some patients it appears to be a useful adjunct to diet in improving glycemic control, and, second, it could retard or prevent the development of macrovascular complications. The evidence for these beliefs will be reviewed as will the notion that exercise may have prophylactic value in young individuals predisposed to type 2 diabetes and/or atherosclerosis.

Glycemic Control. Numerous studies suggest that insulin resistance, in part due to obesity, is often present in type 2 diabetics. That exercise might be useful in coping with such insulin resistance was first suggested by the observation of Bjorntorp et al. [8] that glucose tolerance and insulin sensitivity are both better in middle-aged nondiabetic athletes than in sedentary individuals matched for age and weight (but not for adiposity). In early studies, Ruderman et al. [42] and Saltin et al. [46] demonstrated that several months of regular exercise produce a modest improvement in glucose tolerance in patients with type 2 and chemical diabetes (impaired glucose tolerance), respectively. Schneider et al. [47] later compared 20 sedentary men with type 2 and 11 nondiabetics matched for age, weight and prior

physical activity before and after 6 weeks of thrice weekly training. Glycosylated hemoglobin (a measure of the average blood glucose concentration
over the preceding 8–12 weeks) decreased significantly in the diabetics
(12.2 ± 0.5 to 10.7 ± 0.4%, $p < 0.02$). Despite this, oral and intravenous
glucose tolerance tests performed 72 h after the last bout of exercise
showed only a modest improvement. On the other hand, both fasting
plasma glucose and oral glucose tolerance were significantly better at 12
than at 72 h after the last bout of exercise in 8 subjects tested at both times.
The differences were most marked in patients in whom fasting plasma
glucose was less than 200 mg/dl. The authors concluded that a physical
training program can improve glycemic control in some patients with type
2 diabetes, but that it does so in great measure due to the cumulative effect
of transient improvements in glucose tolerance following each individual
bout of exercise. They also suggested that patients with milder forms of
type 2 diabetes are most likely to show benefits. Improvement in glucose
tolerance and/or glycosylated hemoglobin has also been reported in patients with type 2 diabetes by other investigators [48]. In the largest study
to date, Holloszy et al. [22] have demonstrated a normalization of glucose
tolerance in middle-aged patients with both impaired glucose tolerance
and mild type 2 diabetes after 1 year of intensive training. Like Schneider
et al. [47], this group concluded that benefits were most likely to occur in
patients with mild diabetes in whom insulin resistance was present and
that improvement was at least due in part to the residual effect of the last
bout of exercise. Holloszy et al. [22] attributed their better results than
those reported by others to the intensity of the training regiment (approximately 35 km/week of running). To what extent the fact that their patients
lost 4–5 kg of body weight and decreased body fatness during the year of
training contributed was not ascertained.

An important concept gained from these studies is that the enhancement by regular exercise of insulin sensitivity and glucose tolerance is
more a function of the last bout of physical activity than of the achieved
maximum aerobic capacity of the patient. In keeping with this notion, it
has recently been shown that the increased insulin sensitivity in well-
trained athletes is lost after as little as 3 days of inactivity [10] (fig. 6) and is
restored by a single bout of exercise [20].

Diet and Exercise. As shown by Bogardus et al. [9], patients with type
2 diabetes who lose weight on a hypocaloric diet show little added
improvement in glycemic control when exercise is added to their thera-

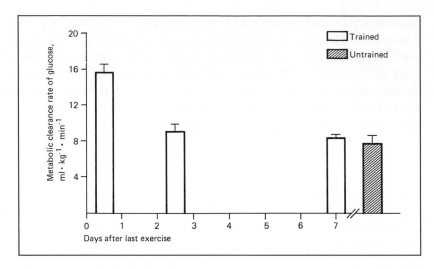

Fig. 6. Effect of detraining on in vivo insulin resistance. Seven aerobically trained athletes (open bars) were studied at 12 h, 60 h, and 7 days after the last exercise period. Three sedentary controls are shown by hatched bars. Results are expressed as mean ± SEM. Adapted from Burstein et al. [10]. Reprinted with permission of the *American Diabetes Association.*

peutic regime. Whether most patients with type 2 diabetes can achieve and maintain the weight loss achieved by the individuals in this study without regular exercise is highly questionable (see discussion of exercise and obesity).

Macrovascular Disease. Atherosclerotic vascular disease affecting arteries in the heart, brain and extremities is accelerated in patients with all forms of diabetes [43]. Epidemiological data, animal studies, and analysis of exercise effects on atherosclerotic risk factors all suggest that physical training retards atherosclerotic vascular disease in the general population [48]. Hypothetically, exercise could be especially useful in the type 2 diabetic as a number of the risk factors affected by exercise are both more prevalent in patients with type 2 diabetes and may contribute to their higher prevalence of macrovascular disease (table II). Investigations of the effect of regular exercise on these risk factors in patients with type 2 diabetes have been limited. Schneider et al. [47] noted a decrease in plasma triglycerides (related to the last bout of exercise) and an increase in the fibrinolytic activity, in 20 patients after a 6-week program of physical

Table II. Possible atherogenic factors in diabetes that could be improved by physical training

Hyperinsulinemia	Abnormal catecholamine response
Hyperlipidermia	to stress
Decreased high density lipoprotein	Hypertension
cholesterol	Obesity
Hypercoagulable state	Tachycardia
Impaired fibrinolysis	Psychological stress

training. Holloszy et al. [22], after a more intense and prolonged (12 months) training program, observed a clear-cut decrease in plasma insulins and an increase in HDL cholesterol.

Exercise and the Prevention of Diabetes. As noted earlier, glucose tolerance is better in middle-aged athletes than in age-matched sedentary controls, suggesting that physical activity may prevent the insulin resistance that contributes to hyperglycemia in people genetically predisposed to type 2 diabetes. In this context, it has recently been found that in Melanesian and Indian men in the Fiji Islands, the prevalence of type 2 diabetes was twice as high in sedentary as in physically active individuals [54]. The difference was observed in both ethnic groups and was still evident when age, obesity, and urban/rural status were taken into account, suggesting that physical activity is an independent risk factor for type 2 diabetes. When a group of urban-dwelling type 2 diabetic Australian Aborigines went on a 7-week 'walkabout' in which they returned to the bush and reverted to a hunter-gatherer lifestyle, glucose tolerance and the insulin response to a glucose load were greatly improved, and the plasma lipid profile was completely normalized [33]. Thus, diabetes appears to be preventable in these people if physical activity, low-fat diets, and control of body weight can be incorporated into their lifestyle. Further evidence in support of this concept comes from a study of 2,622 female former college athletes and 2,776 female nonathletes, ages 20–70+ years [14]. The prevalence of type 2 diabetes was significantly lower in the former athletes compared with the nonathletes for all the age groups studied. The former athletes were also leaner than the nonathletes at all ages up to 70 years, and 74% of the athletes were still actively exercising. These results in total suggest that the risk of developing type 2 diabetes may be markedly reduced by a combination of weight control and a program of regular physical exercise.

Table III. Potential complications of exercise in diabetes

Hypoglycemia	Foot ulcers
Hyperglycemia and ketosis	Orthopedic injury
Retinal hemorrhage	Acceleration of degenerative arthritis
Increased proteinuria	Sequelae of occult ischemic heart disease,
Dehydration and hypotension	e.g. arrhythmias, angina

Exercise in Patients with Diabetes: Practical Considerations

The well-controlled diabetic patient who wishes to exercise should probably be encouraged to do so as part of the overall therapy for his or her disease and for the maintenance of quality of life. Both type 1 and type 2 diabetics should be made aware of the specific problems that exercise might cause if they have significant neuropathy, retinopathy, or other diabetic complications (table III). Thus, intense exercise should be restricted in patients with proliferative retinopathy due to the risk of vitreous hemorrhage. In particular, isometric exercise because it causes a marked increase in blood pressure should be avoided. Jogging and other forms of exercise should be discouraged in patients with peripheral neuropathy as loss of pain sensation, anhidrosis, deformities, and poor circulation make their feet more vulnerable to trauma. Appropriate footwear and frequent inspection of the feet are mandatory in such patients. In these individuals as well as those with neuropathic joints due to diabetes, exercises such as swimming and cycling are preferred. Coronary heart disease is another potential problem in some diabetics. Because of the frequent presence of unrecognized ischemic heart disease [43], a thorough cardiovascular evaluation should be required for all previously sedentary diabetics over the age of 35 about to embark on an exercise program.

Less well-controlled patients are particularly prone to dehydration when exercising on warm days and this predisposes them to postexercise hypotension. While autonomic neuropathy may be a contributing factor, hypotension is most often seen in type 2 patients who are just embarking on an exercise program, and is not a problem once they have trained. Adequate hydration, avoidance of exercise when glycemic control is poor, and a postexercise cooling-down period may be useful in diminishing the problem.

In summary, most patients with diabetes mellitus can engage in recreational physical activity with proper precautions and a good understanding of potential problems. The key to safe and beneficial exercise is adequate patient education and individual monitoring of the glycemic response to physical activity. The role of exercise in the therapy of insulin-dependent diabetes is limited; however, in patients with type 2 diabetes and glucose intolerance, and in young individuals predisposed to diabetes, its role remains to be determined. In patients with significant vascular and/or neurologic complications, the physician must carefully weigh whether the benefits of exercise exceed the risks and if certain types of exercise are preferred.

Osteoporosis

Osteoporosis is a common condition characterized by a loss of bone mass that may lead to fractures and other crippling symptoms. It occurs with greatest prevalence in postmenopausal women. Both sexes lose bone from about age 35 onwards, but the loss is particularly great in women who may lose bone mass at a rate of 2% per year due to diminished estrogen following menopause [19].

An increased bone mass has been reported in athletes and individuals who engage in physically strenuous work [2, 60]. The common characteristic of these activities associated with increased bone mass, i.e. running, ballet dancing, or weight lifting, is that the activity is conducted in opposition to gravity. Thus, increased bone mass is not found in swimmers [32], nor is isometric exercise in bed able to offset the deleterious effects of bed rest on bone mass [60]. The general consensus is that exercise involving weight-bearing movement or at least a vigorous pull of muscle on bone diminishes the rate of bone loss with aging by increasing bone formation [60]. It has been reported, for example, that bone mass is greater in post-menopausal women who are physically active, than in sedentary controls [36].

The mechanism by which exercise stimulates bone formation is uncertain. Bone remodeling is a process by which old bone is removed (resorption) and replaced with an equal amount of newly formed bone. Mechanical forces exerted on bone by muscle activity stimulate the formation of new bone, perhaps by the generation of electrical potentials (piezoelectricity) which activate osteogenesis [5]. Calcium administration appears to diminish the rate of bone resorption [19]. It is important to note that, while exercise stimulates the formation of the new collagen matrix of remodeled bone, the mineralization of the newly formed matrix depends on an adequate supply of calcium. Thus, the two mechanisms are not mutually exclusive. Whatever the exact mechanism, the observation that tennis players have a greater bone mass in their playing than in their nonplaying arms [23, 26] suggests that local factors are involved.

As elderly people tend to be less active physically, it is unclear to what extent this contributes to their increase in osteoporosis. Trained athletes in general have less osteoporosis than sedentary individuals; however, what little evidence is available suggests that they too lose bone mass with age [19]. Presumably, osteoporosis is less marked because they have a larger bone mass to begin with. Nevertheless, such data are consistent with the notion that regular exercise may delay or even prevent the advent of clinically significant bone disorders due to osteoporosis.

The role of exercise in treating established osteoporosis is still under study. An increase of 2.5% in total body calcium has been reported in women 5 years postmenopause who exercised 3 times a week for a year, while a 2.4% loss of calcium was experienced by controls who did not exercise [3]. Similarly, bone mineral content increased by 4.2% in a group of 82-year-old subjects who performed light-to-moderate exercise for 36 months [52]. Exercise is not without risks in these populations, however, since preexisting microfractures can be aggravated by frequent repeated weight-bearing exercise. Inadequate time for repair of the microdamage results in a weakening of the bone.

Obesity

Obesity is defined as an excessive accumulation of body fat to the extent that health is endangered [45]. At present, it is not clear to what extent the adverse effects of obesity are due to increased fat per se, rather than to the other metabolic disorders associated with obesity, such as non-insulin-dependent (type 2) diabetes, hypertension, hypercholesterolemia, hypertriglyceridemia, and hyperinsulinemia [45]. Obesity certainly cannot be viewed as a single disease, but must be evaluated as a complex of metabolic disorders having genetic, endocrine, and environmental factors in its etiology [45].

Whatever the causal mechanism, obesity is the result of an imbalance between caloric intake and expenditure. When caloric intake exceeds expenditure, the excess energy is stored as triglycerides in adipose tissue.

Although overeating is generally viewed as the primary cause of obesity, an increasing body of evidence suggests that differences in the efficiency of energy usage may play a role in many patients [12]. The primary therapy for treating and preventing obesity continues to be a reduction in caloric intake. Interest in the role of exercise has focused on its use as an adjunct to diet.

In addition to the increase in energy expenditure that accompanies physical activity, exercise appears to enhance weight loss in at least two other ways. First, regularly performed exercise maintains the resting metabolic rate. One of the adaptive mechanisms employed by the body in the face of caloric deprivation is a reduction in the resting metabolic rate in order to conserve energy [4]. This response is counterproductive when caloric restriction is used to promote weight loss. It has recently been shown that resting metabolic rate decreased in obese subjects as expected in response to an average 1,500-kcal diet [30]. When a moderate intensity exercise program was added to the restricted caloric intake, the resting metabolic rate was quickly restored to pre-diet levels. There is also evidence that a combined program of caloric restriction and exercise training will result in a relatively greater loss of fat and a sparing of lean tissue than that with caloric restriction alone [34].

A second and equally interesting mechanism by which exercise may enhance weight loss is by the potentiation of dietary-induced thermogenesis (DIT; i.e. the thermic effect of food). DIT is a process by which excess energy is dissipated as heat rather than stored as fat. It is reflected simply as a rise in the metabolic rate following a meal. DIT consists of two components, obligatory thermogenesis and facultative thermogenesis. The obligatory component represents the energy cost associated with the processing and storage of nutrients, while facultative thermogenesis represents the energy expenditure over and above the obligatory component. Available evidence suggests that obligatory thermogenesis is affected most by exercise. The main feature of this component of DIT is the insulin-mediated storage of glucose as glycogen in muscle, an energy-requiring process [37, 38].

The effects of exercise on DIT have been studied by feeding a meal either before or after a bout of exercise, and determining the additional energy expenditure over that observed with either the meal or exercise alone [51]. The results of these studies have been mixed. In those studies failing to demonstrate an enhanced DIT with exercise, the exercise tended to be of mild intensity and short duration. On the other hand, when strenuous exercise was performed, DIT was potentiated. For example, when subjects ingested a carbohydrate meal 2 h after a 45-min bout of exercise at 70% VO_{2max}, by which time postexercise oxygen consumption had returned to pre-exercise levels, DIT was increased by 50% over that observed in response to a meal without prior exercise [61]. No effect of exercise on DIT was found, however, when this protocol was repeated with prior exercise at either 33 or 55% VO_{2max} [55]. These results suggest that DIT is potentiated by exercise when the exercise is of sufficient intensity and duration to reduce muscle glycogen stores.

It has recently been proposed that obesity may, to a certain extent, result from a generalized state of insulin resistance, which in turn decreases the insulin-stimulated component of DIT [12]. This may result in an increased efficiency of weight gain and serve to perpetuate both the obese and the insulin-resistant states. Evidence for this hypothesis comes from the finding that the thermic response to a mixed or carbohydrate meal, normalized for lean body mass, is reduced in obese subjects compared with lean controls [50]. The reduction in DIT is highly correlated with the degree of insulin resistance in these obese individuals.

Exercise training is associated with an increase in peripheral insulin sensitivity. It must be noted, however, that the increase in insulin sensitivity observed in trained individuals is really an acute effect of the last bout of exercise, suggesting the necessity of exercising on a regular basis [20, 47]. Also, exercise must be of sufficient intensity and duration to reduce muscle glycogen stores in order to evoke this response (i.e. 70% VO_{2max} for 40–60 min). This level of exercise may be difficult to attain in very obese individuals. An exercise program may be more appropriate for these individuals after a significant amount of weight has been lost, at which time the level of physical activity can be increased. Finally, because body fat content is inversely correlated with DIT, exercise is an effective way to maintain weight loss once it has been achieved.

Hyperlipidemia

Lipid metabolism is covered elsewhere in the volume, as, to some extent, have lipid changes induced by exercise in individuals with type 2 diabetes and glucose intolerance. This section will therefore be confined to a brief description of exercise effects on plasma triglycerides.

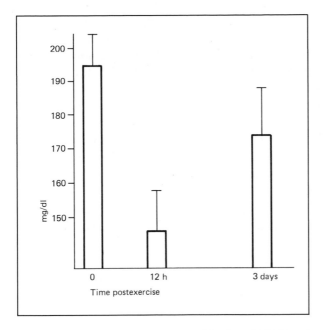

Fig. 7. Effect of a single exercise bout (30 min at 70 % VO_{2max}) on plasma triglyceride levels in 8 patients with type 2 diabetes. Results are expressed as mean ± SEM. Adapted from Schneider and Hanj [48]. Reprinted with permission of John Wiley & Sons, Inc.

Triglycerides, which are transported predominantly in very low density lipoproteins, are more responsive to exercise than is cholesterol. Thus, plasma triglyceride levels are lower in endurance-trained subjects compared with sedentary controls. The degree to which triglycerides decrease in response to exercise depends on the initial plasma levels and the duration of the exercise. In well-trained subjects, who had relatively low triglyceride levels, moderate duration exercise was ineffective in further lowering triglycerides, whereas hypertriglyceridemic patients experienced a significant decrease in plasma triglycerides with exercise of moderate duration [18]. The reductions in plasma triglycerides with exercise, as with glucose, appear to be due to the residual effects of the last bout of exercise, rather than to an exercise-induced adaptation. Plasma triglyceride levels which are significantly reduced 12 h after an acute exercise bout return to baseline levels within 48–72 h (fig. 7) [48]. The reductions in triglycerides are cumulative, however, so that plasma triglycerides decreased from

235 mg/dl initially to 173, 135, 119, and 104 mg/dl on 4 successive days on which male subjects jogged 3–4 miles at a moderate pace [35].

Hyperlipidemia is frequently present in type 2 diabetes. These patients often have elevated levels of triglycerides and cholesterol (if obese) and a reduced high density lipoprotein cholesterol. Studies of type 2 diabetics undergoing a physical training regimen of moderate intensity for 6 weeks report a significant lowering in triglyceride levels, while cholesterol and high density lipoprotein were not affected [42]. As in nondiabetics, the decrease in triglycerides is a function of the last bout of exercise [49].

In summary, the popularity of exercise has increased interest in its use as a therapeutic tool in some metabolic disorders. For certain diabetics and patients with osteoporosis, hyperlipidemia or obesity, exercise may prove to be of therapeutic use in conjunction with appropriate dietary interventions. Research in these areas has increased dramatically in the past 5–10 years, and will undoubtedly continue to do so in the foreseeable future. In particular, the role of exercise in the prevention as well as therapy of these disorders requires careful consideration.

References

1   Ahlborg, G.; Felig, P.; Hagenfeldt, L.; Hendler, R.; Wahren, J.: Substrate turnover during prolonged exercise in man. J. clin. Invest. *53:* 1080–1090 (1974).

2   Aloia, J.F.: Exercise and skeletal health. J. Am. geriat. Soc. *29:* 104–107 (1981).

3   Aloia, J.F.; Cohn, S.H.; Ostuni, J.A.; Cane, R.; Ellis, K.: Prevention of involutional bone loss by exercise. Ann. intern. Med. *89:* 356–358 (1978).

4   Apfelbaum, M.: Adaption to changes in caloric intake. Prog. Fd Nutr. Sci. *2:* 543–559 (1978).

5   Bassett, C.A.L.; Becker, R.O.: Generation of electrical potentials by bone in response to mechanical stress. Science *137:* 1063–1064 (1962).

6   Berger, M.; Berchtold, P.; Cuppers, H.J.; Drost, H.; Kley, H.K.; Muller, W.A.; Wiegelmann, W.; Zimmermann-Telschow, H.; Gries, F.A.; Kruskemper, H.L.; Zimmermann, H.: Metabolic and hormonal effects of muscular exercise in juvenile type diabetics. Diabetologia *13:* 355–365 (1977).

7   Berger, M.; Hagg, S.; Ruderman, N.B.: Glucose metabolism in perfused skeletal muscle. Interaction of insulin and exercise on glucose uptake. Biochem. J. *146:* 231–238 (1975).

8   Bjorntorp, P.; Fahlen, M.; Grimby, G.; Gustafson, A.; Holm, J.; Renstrom, P.; Schersten, T.: Carbohydrate and lipid metabolism in middle-aged physically well trained men. Metabolism *21:* 1037–1044 (1972).

9   Bogardus, C.; Ravussin, E.; Robbins, D.C.; Wolfe, R.R.; Horton, E.S.; Simms, E.A.H.: Effect of physical training and diet therapy on carbohydrate metabolism in patients with glucose intolerance and non-insulin-dependent diabetes mellitus. Diabetes *33:* 311–318 (1984).

10 Burstein, R.; Polychronakos, C.; Toews, C.J.; MacDougall, J.D.; Guyda, H.J.; Pos-
    ner, B.I.: Acute reversal of the enhanced insulin action in trained athletes. Diabetes
    34: 756–760 (1985).

11 Cryer, P.E.; Gerich, J.E.: Relevance of glucose counterregulatory systems to patients
    with diabetes. Critical roles of glucagon and epinephrine. Diabetes Care *6:* 95–99
    (1983).

12 Felig, P.: Insulin is the mediator of feeding-related thermogenesis: insulin resistance
    and/or deficiency results in a thermogenic defect which contributes to the pathogen-
    esis of obesity. Clin. Physiol. *4:* 267–273 (1984).

13 Ferrannini, E.; Barrett, E.J.; Bevilacqua, S.; De Fronzo, R.A.: Effect of fatty acids on
    glucose production and utilization in man. J. clin. Invest. *72:* 1737–1747 (1983).

14 Frisch, R.E.; Wyshak, G.; Albright, T.E.; Albright, N.L.; Schiff, L.: Lower prevalence
    of diabetes in female former college athletes compared with nonathletes. Diabetes
    35: 1101–1105 (1986).

15 Frontera, W.R.; Adams, R.P.: Endurance exercise. Normal physiology and limita-
    tion imposed by pathological processes. Part 2. Physician Sportsmed. *14:* 109–120
    (1986).

16 Galbo, H.; Christiansen, N.J.; Holst, J.J.: Catecholamines and pancreatic hormones
    during autonomic blockade in exercising man. Acta physiol. scand. *101:* 428–437
    (1977).

17 Garetto, L.P.; Richter, E.A.; Goodman, M.G.; Ruderman, N.B.: Enhanced muscle
    glucose metabolism after exercise in the rat. The two phases. Am. J. Physiol. *246:*
    E471–E475 (1984).

18 Haskell, W.L.: The influence of exercise on the concentrations of triglyceride and
    cholesterol in human plasma; in Terjung, Exercise and sports sciences reviews, vol.
    12, pp. 205–244 (Collamore Press, Lexington 1984).

19 Heaney, R.B.: The role of diet and activity in the treatment of osteoporosis; in
    White, Mondeika, Diet and exercise: synergism in health maintenance, pp. 153–159
    (American Medical Association, Chicago 1982).

20 Heath, G.W.; Gavin, J.R.; Hinderliter, J.M.; Hagberg, J.M.; Bloomfield, S.A.; Hol-
    loszy, J.O.: Effects of exercise and lack of exercise on glucose tolerance and insulin
    sensitivity. J. appl. Physiol. *55:* 512–517 (1983).

21 Hilsted, J.; Madsbads, S.; Krarup, T.; Sestoft, L.; Christiensen, N.J.; Tronier, B.;
    Galbo, H.: Hormonal, metabolic, and cardiovascular responses to hypoglycemia in
    diabetic autonomic neuropathy. Diabetes *30:* 626–633 (1981).

22 Holloszy, J.O.; Schultz, J.; Kusnierkiewicz, J.; Hagberg, J.M.; Ehsani, A.A.: Effects
    of exercise on glucose tolerance and insulin resistance. Acta med. scand. *220:* suppl.
    711, pp. 55–65 (1986).

23 Huddleston, A.L.; Rockwell, D.; Kulund, D.N.: Bone mass in lifetime tennis ath-
    letes. J. Am. med. Ass. *244:* 1107–1111 (1980).

24 Hultman, E.; Sjoholm, H.: Substrate availability; in Knuttgen, Vogel, Poortmans,
    Biochemistry of exercise, pp. 63–75 (Human Kinetics, Champaign 1983).

25 Issekutz, B.; Paul, P.: Intramuscular energy sources in exercising normal and pan-
    createctomized dogs. Am. J. Physiol. *215:* 197–204 (1968).

26 Jones, H.H.; Priest, J.D.; Hayes, W.C.: Human hypertrophy in response to exercise.
    J. Bone Jt Surg. *59:* A204–A208 (1977).

27 Kemmer, F.W.; Berchtold, P.; Berger, M.; Starke, A.; Cuppers, H.J.; Gries, F.A.;

Zimmermann, H.: Exercise-induced fall of blood glucose in insulin-treated diabetics unrelated to alterations in insulin mobilization. Diabetes 28: 1131–1137 (1979).

28 Kemmer, F.W.; Berger, M.: Exercise and diabetes mellitus. Physical activity as a part of daily life and its role in the treatment of diabetic patients. Int. J. Sports Med. 2: 77–88 (1983).

29 Koivisto, V.A.; Felig, P.: Effect of leg exercise on insulin absorption in diabetic patients. New Engl. J. Med. 298: 79–83 (1978).

30 Lennon, D.; Nagle, F.; Stratman, F.; Shrago, E.; Dennis, S.: Diet and exercise training effects on resting metabolic rate. Int. J. Obes. 9: 39–47 (1985).

31 National Diabetes Data Group: Classification and diagnosis of diabetes mellitus and other categories of glucose intolerance. Diabetes 28: 1039–1057 (1979).

32 Nilsson, B.; Westlin, N.E.: Bone density in athletes. Clin. Orthop. 77: 179–182 (1971).

33 O'Dea, K.: Marked improvement in carbohydrate and lipid metabolism in diabetic Australian Aborigines after temporary reversion to traditional lifestyle. Diabetes 33: 596–603 (1984).

34 Oscal, L.B.: The role of exercise in weight control; in Wilmore, Exercise and sports sciences reviews, vol. 1, pp. 103–123 (Academic Press, New York 1973).

35 Oscal, L.B.; Patterson, J.A.; Bogard, D.L.; Beck, R.J.; Rothermel, B.L.: Normalization of serum triglycerides and lipoprotein electrophoretic patterns by exercise. Am. J. Cardiol. 30: 775–780 (1972).

36 Oyster, N.; Morton, M.; Linnell, S.: Physical activity and osteoporosis in postmenopausal women. Med. Sci. Sports Exer. 16: 44–50 (1984).

37 Ravussin, E.; Bogardus, C.: Thermogenic response to insulin and glucose infusion in man. A model to evaluate the different components of the thermic effect of carbohydrate. Life Sci. 31: 2011–2018 (1982).

38 Ravussin, E.; Bogardus, C.; Schwartz, R.S.; Robbins, D.C.; Wolfe, R.R.; Horton, E.S.; Danforth, E., Jr.; Sims, E.A.H.: Thermic effect of infused glucose and insulin in man: Decreased response with increased insulin resistance in obesity and noninsulin-dependent diabetes mellitus. J. clin. Invest. 72: 893–902 (1983).

39 Richter, E.A.; Galbo, H.; Holst, J.J.; Sonne, B.: Significance of glucagon for insulin secretion and hepatic glycogenolysis during exercise in rats. Hormone metabol. Res. 13: 323–326 (1981).

40 Richter, E.A.; Garetto, L.P.; Goodman, M.G.; Ruderman, N.B.: Muscle glucose metabolism following exercise in the rat. Increased sensitivity to insulin. J. clin. Invest. 69: 785–793 (1982).

41 Richter, E.A.; Ruderman, N.B.; Schneider, S.H.: Diabetes and exercise. Am. J. Med. 70: 201–209 (1981).

42 Ruderman, N.B.; Ganda, O.P.; Johansen, K.: The effect of physical training on glucose tolerance and plasma lipids in maturity-onset diabetes. Diabetes 28: suppl. 1, pp. 89–92 (1979).

43 Ruderman, N.B.; Haudenschild, C.: Diabetes as an atherogenic factor. Prog. cardiovasc. Dis. 26: 373–412 (1984).

44 Ruderman, N.B.; Young, J.C.; Schneider, S.H.: Exercise as a therapeutic tool in the type I diabetic. Pract. Cardiol. 10: 143–153 (1984).

45 Salans, L.B.: The obesities; in Felig, Baxter, Broadus, Frohman, Endocrinology and metabolism, pp. 891–916 (McGraw-Hill, New York 1981).

46   Saltin, B.; Lingarde, F.; Houston, M.; Harlin, R.; Nygaard, E.; Gad, P.: Physical training and glucose tolerance in middle-aged men with clinical diabetes. Diabetes *28:* suppl. 1, pp. 30–32 (1979).
47   Schneider, S.H.; Amorosa, L.F.; Khachadurian, A.K.; Ruderman, N.B.: Studies on the mechanism of improved glucose control during exercise in type 2 (non-insulin dependent) diabetes. Diabetologia *26:* 355–360 (1984).
48   Schneider, S.H.; Hanj, H.: Clinical aspects of exercise and diabetes mellitus; in Winick, Nutrition and exercise, pp. 145–182 (Wiley, New York 1986).
49   Schneider, S.H.; Vitug, A.; Ruderman, N.B.: Atherosclerosis and physical activity. Diabetes/Metab. Rev. *1:* 513–553 (1986).
50   Segal, K.R.; Guten, B.; Albu, J.; Pi-Sunyer, F.X.: Thermic effects of food and exercise in lean and obese men of similar lean body mass. Am. J. Physiol. *252:* E110–E117 (1987).
51   Segal, K.R.; Pi-Sunyer, F.X.: Exercise, resting metabolic rate, and thermogenesis. Diabetes/Metab. Rev. *2:* 19–34 (1986).
52   Smith, E.L.; Reddan, W.: Physical activity. A modality for bone accretion in the aged. Am. J. Roentg. Rad. Ther. nucl. Med. *126:* 1297 (1978).
53   Susstrunk, H.; Morell, B.; Ziegler, W.H.; Froesch, E.R.: Insulin absorption from the abdomen and thigh in healthy subjects during rest and exercise. Blood glucose, plasma insulin, growth hormone, adrenaline and noradrenaline levels. Diabetologia *22:* 171–174 (1982).
54   Taylor, R.; Ram, P.; Zimmet, P.; Raper, L.R.; Ringrose, H.: Physical activity and prevalence of diabetes in Melanesian and Indian men in Fiji. Diabetologia *27:* 578–582 (1984).
55   Treadway, J.L.; Young, J.C.: The effect of prior exercise upon the thermic effect of glucose. Dependence on exercise intensity. Diabetes *35:* suppl. 1, p. 742 (1986).
56   Vranic, M.; Berger, M.: Exercise and diabetes mellitus. Diabetes *28:* 147–163 (1979).
57   Vranic, M.; Karwamori, R.: Essential roles of insulin and glucagon in regulating glucose fluxes during exercise in dogs. Diabetes *28:* suppl. 1, pp. 45–52 (1979).
58   Wallberg-Henriksson, H.; Holloszy, J.O.: Activation of glucose transport in diabetic muscle: responses to contraction and insulin. Am. J. Physiol. *249:* C233–C237 (1985).
59   Wasserman, D.H.; Lickley, H.L.; Vranic, M.: Interactions between glucagon and other counterregulatory hormones during normoglycemic and hypoglycemic exercise in dogs. J. clin. Invest. *74:* 1404–1413 (1984).
60   Whedon, G.D.: Interrelation of physical activity and nutrition on bone mass; in White, Mondeika, Diet and exercise: synergism in health maintenance, pp. 99–112 (American Medical Association, Chicago 1982).
61   Young, J.C.; Treadway, J.L.; Balon, T.W.; Gavras, H.P.; Ruderman, N.B.: Prior exercise potentiates the thermic effect of a carbohydrate load. Metabolism *35:* 1048–1053 (1986).
62   Zinman, B.; Vranic, M.; Albisser, A.M.; Leibel, B.S.; Marliss, E.B.: The role of insulin in the metabolic response to exercise in diabetic man. Diabetes *28:* suppl. 1, pp. 76–81 (1979).

John C. Young, MD, Department of Health Sciences, Sargent College,
Boston University, Boston, MA 02215 (USA)

Subject Index